AIR TO AIR MISSILE
DIRECTORY

Hugh Harkins

Air to Air Missile Directory

A Centurion Book

© Hugh Harkins 2014

ISBN 10: 1-903630-25-8
ISBN 13: 978-1-903630-25-9

This volume first published in 2014

Published by Centurion Publishing, United Kingdom

Cover design © Centurion Publishing & Createspace

Page layout, concept and design © Centurion Publishing

A UAE Block 60 F-16 fighter armed with AIM-120 AMRAAM missiles on the wingtip stations. LMTAS

Air to Air Missile Directory

Contents

Introduction

In the 21st Century air combat is dominated by the world of advanced electronics and surveillance systems including advanced radar, either fire control and search, or large surveillance systems, which can be surface based or mounted on airborne platforms such as the Boeing E-3 AWACS (Airborne Warning and Control System). These systems have proven to be force multipliers and battle winners in recent air campaigns where United States and other NATO nation's fighter aircraft have been put in positions of advantage when engaging opponents which lacked such advanced systems. However, in the end game of an engagement the fighter aircraft must posses advanced missiles to take advantage of the force multipliers afforded by the AWACS and other systems. Since the late 1990's and into the 21st Century a number of new advanced air to air missiles have either entered service or are in development, while older systems like the Raytheon AIM-9 Sidewinder and AIM-7 Sparrow have remained in service. The variants of these latter weapons in current service are a world apart from the original systems fielded in the 1950's and 1960's.

In Russia, older air to air missiles like the R-4 AA-5 'Ash' used in the large Tupolev Tu-128 'Fiddler' interceptor were retired along with the launch aircraft at the end of the Cold War in the early 1990's. The K-8 (R-8/98) AA-3 'Anab' was also retired when the Sukhoi Su-15 and Yakovlev Yak-28P 'Brewer' interceptors were retired at the end of the Cold War. The R-8M (R-98) equipped with a SARH (Semi-Active Radar Homing) seeker was capable of engaging targets head-on as well as from the traditional tail chase position behind the target. This missile was used on the Su-15-98/Su-15TM and Yak-28P. The Su-15TM later employed the R-98M with enhanced resistance to countermeasures from 1973. The R-8/98 had a range of 19-km and a speed of Mach 3. It was the R-98 that Soviet Su-15 'Flagon' interceptors used to shoot down the Boeing 747 KAL003 of Korean Air Lines, which had strayed over Soviet Air Space in 1983.

In the 1980's, the then Soviet Union introduced a whole family of new air to air weapons including the R-33 long-range SARH and medium-range R-27R SARH and R-27T IR (Infrared) homing missiles. While these weapons were considered to be roughly equivalent to the US AIM-54C Phoenix long-range missile and the AIM-7F/M medium-range SARH missiles, the Soviet Unions R-73 short-range IR guided missile, which appeared in the mid-1980's was considered to be far superior to the best NATO equivalent, the AIM-9M Sidewinder. Adding to NATO's headache was the fact that the latest generation of Soviet agile fighters like the Su-27 and MiG-29 were equipped with a simple, but effective, HMS (Helmet Mounted Sight), which allowed the R-73 to be cued at high off-boresight angles, increasing the already impressive dogfight capability of the Soviet fighters.

While Russian and Chinese air to air missiles continue to sell to nations purchasing Russian and Chinese fighters, the western air to air missile market has in recent years been dominated by the US Raytheon AIM-120 AMRAAM and derivatives and the European consortium Meteor for the active-radar guided BVR (Beyond Visual Range) role. The US Raytheon AIM-9X, MBDA ASRAAM and the BGT IRIS-T have dominated the western and western customer nations short-range air to air missile markets.

In both markets the European manufacturers have got their work cut out for them if they are to remove Raytheon from the top spot in air-to-air missile manufacturing. From the early 2000's, European nations have been taking delivery of around 640 Eurofighter Typhoon, 294 Dassault Rafale and just over 210 Saab JAS 39 Gripen fighters, the later entering service in 1997. This combined total of 944 fighters to be armed with Meteor will pale in comparison to the planned US purchase of some 200 F-22 Raptors, in excess of 2,000 F-35 JSF's and just under 500 F/A-18E/F Super Hornet strike fighters, all to be armed with AMRAAM (The numbers of the F-22 and F-35 have been altered several times and may be subject to further alterations). Added to this are the thousands of third generation Boeing F-15, Lockheed Martin F-16 and Boeing F/A-18A/B/C/D Hornets already armed with AMRAAM as their primary air-to-air armament. AMRAAM's dominance is further increased by the fact that Sweden's JAS 39 Gripen and British German, Italian and Spanish Eurofighter Typhoons were delivered with AMRAAM as their primary air to air weapon as those nations' awaited development and delivery of the pan-European Meteor. The RAF's Panavia Tornado F.3 CSP (Capability Sustainment Program) fighters, Germany's F-4F Phantom II ICE (Improved Combat Efficiency) and Swedish Saab Viggen fighters were all modified to operate with AMRAAM. Europe's only counter to AMRAAM in the shorter term was the MBDA MICA, which is available in both active-radar and infrared homing variants. MICA however was in a poor position to compete with AMRAAM as it had only been integrated onto Dassault Rafale and Mirage 2000-5/9 Series fighters.

MICA has been credited with a range of around 60-km, however, the design was not considered to be capable of extending the range required for the EBVR (Extended Beyond Visual Range) mission studied by a number of air forces, including that of France. The EBVR was required to counter extended range BVR Russian missiles in service or under development and would also allow non-stealth driven fighters such as the Saab Gripen, Eurofighter Typhoon and Dassault Rafale to remain outside an opponent's missile engagement zone during an air to air engagement. This type of weapon would almost certainly require rocket/ramjet propulsion and MICA would have had to undergo a complete redesign in order to achieve this. For this reason France joined the pan-European Meteor EBVR missile program with the aim of enhancing the air-to-air capability of the Rafale.

With the Lockheed Martin F-35 Lightning II expected to dominate the export fighter market in the next few decades, Meteor may already have lost the competition with AMRAAM and its derivatives despite Meteors obvious superiority over its US rival. The F-35 can carry the AMRAAM in its internal weapon bays; however, during the design phase of the F-35 program it was decided not to include internal carriage of Meteor into the design. This may have been a clever ploy by the US to ensure domination of the BVR air-to-air missile market and a spectacular own goal by the European partners in the F-35 program.

In the IR guided missile arena European manufacturers are divided with two main programs - ASRAAM and IRIS-T - competing against the US AIM-9X. The MBDA MICA IR (Infrared) variant is unlikely to be a competing factor for most competitions other than those effecting Mirage 2000 or Rafale fighters. On the plus side for the European manufacturers is the fact that ASRAAM has been launched from a Lockheed Martin F-16 Fighting Falcon and purchased by Australia, which operates the missile with its upgraded Boeing F/A-18A/B Hornet strike fighters. The IRIS-T is also to be integrated with the F/A-18C/D Hornet. This will mean future export customers will not have to pay for integration of the missile with either of these types. However, the low unit cost of the AIM-9X compared with the European competitors has seen the AIM-9X ordered by a number of European air arms. Switzerland has ordered the missile to arm its F/A-18C/D Hornets and Denmark has ordered the AIM-9X to arm its upgraded F-16AM/BM Fighting Falcons.

While the European Airbus Industries eventually knocked the US Boeing Company from the top spot in Airliner manufacture and sales, even if only temporarily, the European military aircraft and missile sectors can only dream of such a position. As long as European air forces continue to purchase large fleets of US combat aircraft and weaponry then the European manufactures will have to continue to settle for second place to the giants of the US military industrial base. While the European Meteor consortium is looking to manufacture a few thousand missiles for no less than seven European air forces, the US is manufacturing well in excess of ten thousand AIM-120 AMRAAM's for the domestic market alone, with thousands more being produced for the export market. The same position will apply to the IR guided missile systems, with the US producing tens of thousands of AIM-9X missiles, while European manufactures produce mere handfuls in comparison.

A number of missile systems have been adapted for use or designed specifically as air to air weapons employed by modern attack helicopters. A variant of the Raytheon AIM-9 Sidewinder has been employed on USMC Bell AH-1W attack helicopters and the Russian R-73 has been integrated with Russian attack helicopters. The Raytheon Stinger MAPAD (Man Portable Air Defence) shoulder launched IR guided surface to air missile has been adapted for launch from attack helicopters. The Shorts Starstreak has been integrated with the Boeing AH-64D Apache Longbow variant, giving this platform an impressive air-to-air self-defence capability. China is developing the Tianyan (Celestial Swallow)-90 to arm the Z-9 attack helicopter and EADS in Europe is developing the LFK NG Le Fla as an air-to-air missile for attack helicopters with the Europcopter Tiger seen as the primary platform. Missile systems designed specifically for helicopter launch have not been included in the following chapters, nor is the AIM-54 Phoenix long range air to air missile which was withdrawn from USN when the F-14 Tomcat was retired in 2007. Iranian F-14's may still operate handfuls of these missiles in the air to air role, however, and there have been a number of sightings of Iranian Tomcats with Phoenix or Phoenix derived missiles, although the operational status of the missiles is unclear.

This volume has been prepared using manufacturers and operator's data.

Chapter One

Over Three Decades of Air Combat

Over the past several decades there have been a number of conflicts which have resulted in considerable air combat activity as well as short engagements which have resulted in aircraft being shot down by either air-to-air missiles or aircraft cannon fire. In the majority of these engagements the nation which operated the most modern combat aircraft armed with modern air-to-air missiles, more often than not, came out on top.

In the years followings its unfavourable withdrawal from Vietnam in 1973, the United States, particularly after it entered the Carter era of the late 1970's, was moving more towards diplomacy to try and resolve problems on the international stage. However, the fall of the Shah of Iran in the Iranian Islamic revolution of 1979, together with the election of Ronald Reagan as United States President set the stage for US military confrontations in the oil rich Middle East, which continue to this day. After the debacle of operation Eagle Claw, the disastrous attempt to rescue US hostages in Iran, the US turned its attention to the Western Mediterranean and began a serious of moves which would, intentionally or not, provoke Libya in an armed confrontation.

Following a number of air incidents which ended without either side firing on the other, the calm between the US and Libya was broken on 19 August 1981 when a pair of USN (United States Navy) Grumman F-14 Tomcat fleet air defence fighters engaged and shot down a pair of Libyan Air Force Su-22 'Fitter' ground attack aircraft over the Gulf of Sidra off the Libyan Mediterranean coast. The USN Tomcats each launched an AIM-9L at the 'Fitters', which were destroyed. Further conflict between Libya and the US erupted in the mid-1980's, which included operation Eldorado Canyon – a short US bombing campaign against Libya in April 1986. There were no recorded air combat engagements during this operation.

On 4 January 1989, a pair of USN Grumman F-14 Tomcats shot down a pair of Libyan MiG-23MS 'Flogger' fighters, which had not fired on the US fighters. An AIM-7M Sparrow was launched at the first MiG-23, but missed. This was followed by another AIM-7M launched at the same aircraft; this time the missile destroyed the MiG. The second MiG-23 was destroyed by an AIM-9M launched from the second F-14. This incident stirred up much controversy as it appeared that the US was deliberately trying to provoke a conflict with Libya. Similar encounters between any US or other NATO (North Atlantic Treaty Organisation) fighters and Soviet aircraft had always resulted in no weapons launch from the NATO fighters. Indeed, there were countless such encounters between Soviet and NATO, or even neutral Swedish aircraft, which passed without incident. However, with the US's attention turning towards the Persian Gulf in summer 1990 there were no further clashes between US and Libyan forces.

The Arab Israeli conflicts had seen much fighting between the Jewish State and its Arab neighbours, fuelled by the constant flow of arms by the Super Powers. Following the Yom Kippur war in 1973, tensions remained high between Israel and its neighbours, particularly Syria to the North. It is claimed that on 27 June 1977, Israeli Air Force McDonnell Douglas (later Boeing) F-15A Eagle fighters shot down 4 Syrian Air Force obsolete MiG-21 'Fishbed' fighters over Lebanon in the first ever combat use of the new F-15 air superiority fighter. The missile types used are unclear, but were probably either AIM-7F BVR (Beyond Visual Range) Sparrow or late model AIM-9 Sidewinders or a mix of both. Another claim is that Israeli F-15A's shot down five Syrian fighters on 24 September 1979 and another on 27 June 1980. It is claimed that an Israeli F-15A shot down a Syrian MiG-25PD with an AIM-7F Sparrow on 13 February 1981. Another Syrian MiG-25PD was claimed by Israeli F-15A's later in 1981 followed by Syrian claims that a MiG-25PD shot down a Israeli F-15A using a pair of R-40's with the IR guided R-40T launched ahead of the SARH R-40R to prevent the IR missile homing on the rocket motor of the R-40R. The R-40's are reported to have been launched from a range of around 40-km (25 miles). It should be noted that many of these Israeli and Syrian claims cannot be reliably verified.

In May 1982, one month before the hectic air battle over the Bekaa Valley in June, it is claimed that Israeli F-15's shot down a pair of Syrian MiG-23's. During the Israeli invasion of Lebanon in June 1982 Israeli F-15's, General Dynamics (later Lockheed Martin) F-16 and IAI Kfir fighters engaged the Syrian Air Force in large scale air to air combat. These engagements were to dramatically show the superiority gained by Israel in the air. This was achieved through a number of factors. Israeli tactics proved vastly superior to the tactics employed by Syria; however, the Israeli dominance of the air over the Beka Valley in June 1982 was attributed more to the modern airpower and weapons employed in the face of Syria's obsolete fighter force. Israel employed the modern F-15 armed with BVR AIM-7 Sparrow missiles, and close range missiles, either all-aspect AIM-9L Sidewinders or the then new Rafael Python 3 all-aspect IR guided air to air missile. The F-16 Fighting Falcons lacked a BVR missile capability, but were armed with the all-aspect AIM-9L. Other Israeli fighters were armed with older, but effective weapons; either older Sidewinder models or indigenous Rafael Shafrir IR guided air to air missiles. However, it was the F-15 and F-16, which took on the lion's share of the air to air engagements with Syrian fighters using AIM-7 Sparrow and AIM-9L Sidewinders.

There are conflicting claims as to how many Syrian Air Force aircraft were actually shot down during the fighting of 5-12 June 1982, however, the figure seems likely to be around 90, mainly obsolete MiG-21 and MiG-23 fighters. At least three MiG-25 'Foxbat' high altitude, high Mach, fighters were claimed by Israel. There are claims that a pair of Syrian Air Force MiG-23's was shot down by Israeli fighters in 1985. A more recent acknowledged air combat engagement between Israel and Syria took place off the Syrian coast in late 2004 when Israeli F-15 Eagles shot down a pair of Syrian MiG-29 'Fulcrums' using AIM-9M Sidewinder IR guided missiles.

In January 1981, during a 13-day border conflict with Peru, an Ecuadorian Air Force Dassault Mirage F.1JA(E) launched a Matra Magic 1 IR guided air-to-air missile at a Peruvian Air Force Sukhoi Su-22 'Fitter'. The missile, which was launched from a wingtip launch rail, was apparently launched outside parameters and failed to achieve a kill.

The stage was set for Britain's most intense period of air combat operations since the end of World War 2 when on 2 April 1982 Argentina invaded the remote British administered Islands of South Georgia and the Falklands in the south Atlantic. This resulted in Britain assembling a task force to retake the Islands. On 1 May 1982 an Argentine Air Force Mirage III fighter launched a Matra R.530 SARH air-to-air missile at a Royal Navy Sea Harrier FRS.1 during the first aerial engagement between the Argentine and British fighters. The missile was launched without a proper radar-lock and failed to guide. This was the closest Argentine fighters would come to success over British Sea Harriers. The British fighters on the other hand went on to shoot down over 20 Argentine aircraft using all the aspect AIM-9L Sidewinder IR guided air-to-air missiles, which were much superior to anything the Argentines had in their inventory. There were no missiles launched from the head-on aspect. However the superior seeker head of the 'Nine Lima' (AIM-9L) hugely increased the combat efficiency of the British fighters.

On 5 October 1982, a South African Air Force Mirage F1CZ shot down an Angolan Air Force MiG-21 'Fishbed' using a Magic 1 IR guided missile. Another South African Air Force fighter shot down an Angolan MiG-23 during the conflict in southern Africa.

During the Iran Iraq war of 1980-88 both sides shot down opposing forces aircraft in air combat. It is known that at least one IRIAF (Islamic Republic of Iran Air Force) Grumman F-14A Tomcat was shot down by an Iraqi Air Force Mirage F1, although there are claims that a up to four F-14's were shot down in air combat.

On 5 June 1984, RSAF (Royal Saudi Air Force) McDonnell Douglas F-15 Eagles engaged and shot down a pair of IRIAF McDonnell Douglas F-4D Phantom II fighters over the Persian Gulf during a period of border and airspace tensions in the region. The Eagle's armament consisted of AIM-7F Sparrow BVR and AIM-9P/L Sidewinder air-to-air missiles. It is unclear which weapons were employed, various reports stating the use of differing weapons.

There are claims that USN F-14 Tomcat fighters launched a number of AIM-7F Sparrow and AIM-9L Sidewinder air-to-air missiles at IRIAF fighters during operations in the Persian Gulf in 1988/89. During one claimed incident, a pair of Tomcat's from VF-21 launched an AIM-7F and an AIM-9L at a pair of Iranian F-4 Phantom II's, with both missiles missing their targets.

While the US appeared to be provoking a fight with Iran in 1988/89, attention switched to Iraq following that nation's invasion of Kuwait on 2 August 1990. Following authorisation from the United Nations a large multi-nation coalition was assembled to force Iraqi forces to leave Kuwait. During the resulting Gulf War of January to March 1991, US and Saudi fighters shot down over 40 Iraqi aircraft using air-to-air missiles and guns. The coalition air forces headed by the US, equipped with the most advanced weapons and sensors available, including ECCM (Electronic Counter Counter Measures) suppression and AWACS (Airborne Warning and Control System), enjoyed complete dominance of the intelligence/surveillance situation in Kuwait and Iraq. The Iraqi's on the other hand were equipped with few modern fighter aircraft with any significant air-to-air capability. Most of the MiG-

21, MiG-23 and Mirage F1 fighters were, by the start of the 1990's, well by their sell by date as was most of the associated weaponry of these aircraft. However, despite having a massive superiority in both numbers of fighter aircraft available and in advanced technology, the coalition air war winner was undoubtedly the AWACS, which allowed the coalition dominance of the Skies over Iraq and Kuwait from the beginning of the air war. Added to this was the fact that the Iraqi Air Force was under constant air and cruise missile attack at its bases in Iraq and Kuwait. Outnumbered and technologically outclassed the Iraqi Air Force, which looked strong on paper, was never going to be capable of successfully challenging the coalition air armada arrayed against it. Before the first shot was fired, this was always going to one of the most one-sided air conflicts in history.

On the first day of the war, 17 January 1991, US fighters shot down eight Iraqi fighters. USAF F-15C's shot down 3 Iraqi Mirage F.1's using AIM-7M Sparrow BVR missiles. F-15C's also shot down three MiG-29's using AIM-7M's. A pair of USN F/A-18C Hornet strikes fighters on an air to surface strike mission engaged and shot down a pair of Iraqi F-7A (Chinese MiG-21 copy) fighters using IR guided AIM-9M Sidewinder air to air missiles. However, it was not all one sided on the first day of the war. In the only confirmed coalition aircraft loss to an Iraqi fighter aircraft an Iraqi Air Force MiG-25 'Foxbat' E fighter engaged and shot down a USN McDonnell Douglas F/A-18C Hornet strike fighter. It remains unclear what type of weapon was used, but this was either an R-60 AA-8 'Aphid' or an R-40R (SARH) or R-40T (IR) AA-6 'Acrid', which were carried by the MiG-25. The only other Iraqi air combat success that day was a MiG-29 forcing a USAF General Dynamics F-111F strike fighter to drop its bomb load and turn for friendly territory to escape.

The next air combats took place on 19 January when USAF F-15C's engaged and shot down two Iraqi MiG-25 'Foxbat' E's using AIM-7M's. F-15C's also shot down a MiG-29 and a pair of Mirage F.1's using AIM-7M's on 19 January and another Iraqi MiG-29 crashed while manoeuvring during the days operations. Iraqi air defences shot down a USN F-14A Tomcat from VF-103 on 21 January 1991. It was speculated that this could have been an air to air combat loss, although it is officially listed as a loss to ground fire. RSAF (Royal Saudi Air Force) F-15C's engaged and shot down a pair of Iraqi Mirage F.1's using AIM-9 Sidewinder IR guided missiles on 24 January. The Mirage F.1's were thought to have been attempting an anti-shipping strike against coalition surface vessels in the Persian Gulf using AM 39 Exocet anti-ship missiles. On 26 January, F-15C's

shot down three Iraqi MiG-23 'Flogger's' using AIM-7M's. The following day, 2 MiG-23's were shot down by USAF F-15C's using AIM-9M Sidewinders. Another MiG-23 and a Mirage F.1 were shot down by F-15C's using AIM-7M's. USAF F-15C's shot down another two MiG-23's on 29 January using AIM-7M's. On 6 February, USAF F-15C's shot down two MiG-21's and two Su-25 ground attack aircraft using AIM-9M's. A USN F-14 Tomcat shot down an Iraqi Army Mi-8 'Hip' transport helicopter using an AIM-9M. The next day, 7 February, four Sukhoi ground attack aircraft and an Iraqi helicopter were shot down by USAF F-15C's using AIM-7M's. Another pair of Iraqi helicopters was shot down by F-15C's using AIM-7M's on 11 February.

A ceasefire between coalition and Iraqi forces was called on 28 February and signed on 3 March 1991. However, as history has shown, the Iraq conflict was only just beginning and would continue into the second decade of the 21st Century.

The next air engagements between coalition and Iraqi forces took place on 20 March 1991, when USAF F-15C's shot down an Iraqi Su-22 'Fitter' ground attack fighter using an AIM-9M Sidewinder. Another Su-22 was shot down by USAF F-15C using an AIM-9M Sidewinder on 22 March 1991.

During the January-March 1991 Gulf War, RAF Jaguars apparently launched three AIM-9L Sidewinders, although this was apparently in error. Canadian Armed Forces CF-18 (CF-188) Hornet strike fighters were called-off a CAP to strafe and Iraqi surface vessel which had been attacked and damaged by coalition attack aircraft. Once out of 20-mm cannon ammunition the Hornet pilots launched AIM-7 Sparrow air-to-air missiles at the vessel, but missed. The vessel was eventually sunk by a fresh wave of strike aircraft.

In the 18 months after the ceasefire of March 1991, Britain, France and the US set up controversial NZF (No Fly Zones) over Northern Iraq above the 36th parallel and over Southern Iraq below the 32nd parallel (later extended to the 33rd parallel. The NFZ's were set up without United Nations sanction and were, therefore, unilateral operations with no specific resolution to authorise them. France later withdrew from the Northern operation and took no part in the large number of British and US air strikes against Iraqi targets from 1998 to 2003.

The months December 1992 and January 1993 saw a number of Iraqi challenges to the US and its coalition partners. Coinciding with the last weeks of the US President George Bush senior's administration, Iraq seamed intent on testing the will and resolve of the United States and her allies to continue enforcing the NFZ's over the 36th Parallel and below the 32nd Parallel. The first air to air incident occurred early on the morning of 27 December 1992 when a pair of Iraqi air force fighters crossed the 36th parallel into the northern NFZ and challenged USAF F-15E Eagles of the 355th Fighter Squadron (FS), 4th Fighter Wing (FW). Later another pair of Iraqi fighters crossed the 32nd parallel into the southern NFZ, but turned back North when challenged by a pair of USAF F-15C Eagles from the 1st FW which were vectored onto the incoming Iraqi fighters by an USAF E-3C Sentry AWACS.

At 07:20 am GMT a pair of Iraqi MiG-25 'Foxbat' E interceptors crossed the 32nd parallel entering the Southern NFZ. The MiG-25's were detected by an USAF E-3C Sentry AWACS, which vectored USAF F-16C/D Block 42 Fighting Falcons of the 33rd FS, 363rd FW, to engage them. The F-16's were operating out of Al Dhafra air base in Sharjah (part of the UAE). There have been conflicting reports of the incident, but as far as can be ascertained, the Iraqi fighters turned north back across the 32nd parallel. Then a second pair of Iraqi fighters, again MiG-25 E's, although some reports claim that they were MiG-29's, crossed into the Southern NFZ at 17:40 GMT. It is claimed that the Iraqi fighters turned onto a heading apparently attempting to engage the American fighters while some 32-km (20 miles) inside the Southern NFZ. The F-16's then engaged the Iraqi fighters. The lead American fighter, an F-16D, fired a single AIM-120 AMRAAM from Beyond Visual Range (BVR) at the lead Iraqi fighter that was now 37-km inside the (NFZ). The missile destroyed the Iraqi fighter and the second MiG-25 then quickly returned to Iraqi controlled air space. This action was significant for the USAF in that it saw the first USAF use of the F-16 to shoot down an enemy aircraft and the first combat use of the Raytheon AIM-120 AMRAAM.

Another incident occurred on 2 January 1993, when a MiG-25 'Foxbat' E attempted to intercept an USAF Lockheed U-2R reconnaissance aircraft, and was subsequently "chased off" by USAF F-15C Eagles. On the 17th, a USAF F-16C Block 30, 86-0262, from 2the 3rd FS, 52nd FW stationed at Incirlik with the 7440th Composite Wing, engaged and shot down an Iraqi Air Force fighter. The Iraqi fighter was widely reported as being a MiG-23 'Flogger', but later identified as a MiG-29 'Fulcrum' which had crossed into the Northern NFZ. The aircraft was, it is claimed, engaged as it attempted to escape south, but Iraqi radio reported that the aircraft was shot down while taking-off from an air base south of the 36th Parallel. Pentagon and USAF officials later claimed that the aircraft was engaged and hit inside the NFZ, but the wreckage had crashed onto the ground just south of the 36th Parallel.

On 18 January an USAF F-15C engaged a MiG-25E, which according to US officials "took an attack posture." The Eagle launched one AIM-120 AMRAAM from a range of 43-km (27 miles), then launched an AIM-9M Sidewinder from closer range. The MiG-25 was reported as being shot down, but later analysis led the US Department of Defence to conclude that the MiG-25 had returned safely to base.

In April 1994, a pair of US Army Sikorsky UH-60 Black hawk helicopters was shot down by a pair of USAF F-15C Eagles of the 52nd FW, Spangdahelm, which had mistaken them for Iraqi machines entering the NFZ. Both Blackhawks were destroyed killing all 26 occupants. After wrongly visually identifying the Blackhawks as Iraqi helicopters the F-15's were ordered to close on the helicopters and after wrongly visually identifying them as Iraqi; they were cleared to engage by the Boeing E-3B Sentry. One F-15 fired an AIM-120A AMRAAM and the other an AIM-9M Sidewinder missile, both hitting the helicopters.

In late December 1998, the US and British coalition of two launched a four day air campaign against Iraq. This set the stage for over three years of limited air strikes against Iraq as Iraq began challenging coalition aircraft flying over Iraqi airspace. The first real air to air engagement between Iraqi and US fighters since 1993 took place on 5 January 1999. In the first of two incidents in the southern NFZ that day, the US claimed that a pair of USAF Boeing F-15C Eagles flying from Saudi Arabia were illuminated by the Fire Fox radar on what was claimed as four Iraqi Air Force MiG-25 'Foxbat' E's. The US fighters responded by firing a single Raytheon AIM-7M Sparrow semi-active radar guided and 3 AIM-120 AMRAAM active radar guided air to air missiles at the Iraqi fighters, but all four of the missiles missed their targets and the Iraqi fighters escaped north.

The second of the day's incidents in the southern NFZ occurred 15 minutes later when a pair of USN Northrop Grumman F-14 Tomcats flying from the *USS Carl Vinson* (CVN 70) in the Arabian Gulf fired a pair of AIM-54C/ECCM/Sealed Phoenix long range air to air missiles at a pair of MiG-25's flying just south of Baghdad, but heading north following an incursion into the NFZ. However, like the earlier incident, the US missiles missed their targets and the Iraqi fighters escaped. It has been reported that the Iraqi aircraft were actually outside the NFZ when the missiles were launched at extreme range.

Altogether there were eight separate reported incidents of Iraq aircraft entering the southern NFZ involving 15 Iraqi fighters on 5 January. Some reports indicated that one of the Iraqi MiG's had crashed after running out of fuel, but this report was denied by Iraq and no tangible evidence to support the claim was forthcoming from the US DoD.

In December 2002, the Pentagon claimed that an Iraqi MiG-25E intercepted and shot down a USAF General Atomics RQ-1A Predator UAV (Uninhabited Air Vehicle), although Iraq claimed that ground based air defences had shot down the Predator. If it was an air interception then again it is not known what missile type was used. If this was an air-to-air loss, then it has the distinction of being the last air-to-air engagement between Iraqi and US aircraft.

The air combat involving western fighter aircraft and missiles in the 1990's proved to be far more successful than previous generations of aircraft and missiles. The modern variants of the Sparrow and Sidewinder fielded by the US in the 1990's were inherently more capable than the earlier generations fielded during the Vietnam War. Between May 1972 and January 1973 the AIM-7 and AIM-9 had a dismal combat record. Only around 10% of AIM-7's launched hit their targets, while the AIM-9 fared only slightly better with around 11% of missiles hitting their targets. During this conflict, the AIM-7 was reluctantly used at BVR as it could just as likely engage friendly aircraft as it could enemy aircraft. By contrast, the technological advances, particularly with the IFF (Identification Friend or Foe) and the airborne control of fighter aircraft with Boeing E-3 AWACS aircraft allowed BVR engagements to be conducted with a much higher degree of reliability. 16 of the 23 AIM-7M air-to-air kills credited in the 1991 Gulf War were conducted from beyond visual range.

While AMRAAM performed poorly during air to air engagements between US and Iraqi fighter aircraft, the USAF achieved better results during operation Allied Force against Serbian aircraft during the March to June NATO bombing campaign against Serbia, Montenegro and Kosovo. USAF F-15 Eagle's shot down 4 Serbian MiG-21/29's using AMRAAM's. Another Mig-29 was shot down by an AMRAAM launched from a RNAF (Royal Netherlands Air Force) F-16AM. Five years earlier, in 1994, USAF F-16C's shot-down four Serbian aircraft over Bosnia using AIM-9M and AIM-120's.

Tensions between Pakistan and India, particularly over the disputed Kashmir region began to heat-up significantly in 1999 and into the early 2000's. Although full scale hostilities have so far been avoided, there have been some heavy fighting involving ground forces from both sides and a number of air incidents with aircraft from both sides being either shot down or engaged by the others

ground based air defences. During a rare air-to-air engagement between the two sides an Indian Air Force MiG-21 'Fishbed' shot down a Pakistan Air Force Dassault Atlantic maritime maritime/reconnaissance aircraft with AA-8 'Aphids'. Two Aphids were launched at the large-patrol aircraft.

The air combat records of western aircraft over the past decade and a half show a clear superiority over their opponents. This is hardly surprising as during the campaigns in the Persian Gulf region and over the Balkans the US and her coalition allies enjoyed an overwhelming numerical and technical superiority over their opponents. The use of E-3 AWACS and other electronic intelligence systems allowed the coalition aircraft to engage their opponents often from BVR and often without their opponents even knowing they were being engaged. The AWACS also gave friendly aircraft advanced warning of any enemy attempts to set up an engagement.

What the few decades have shown is that not only has air-to-air missile technology advanced to a stage where it is considered very reliable, the use of advanced sensors on tactical fighters combined with force multipliers such as AWACS enable complete superiority over an adversary lacking such assets.

European Air to Air Missiles

MBDA ASRAAM

For short range air to air combat the MBDA (Matra BAc Dynamics Alenia) ASRAAM (Advanced Short Range Air to Air Missile) is employed by RAF Eurofighter Typhoons, initially T.1/ F.2's and later T.3 and FGR.4's. The missile also armed Panavia Tornado F.3 Interceptors and Boeing BAE Systems Harrier GR.7/9 in RAF service prior to the retirements of these types in 2011 and 2010 respectively.

ASRAAM was designed to replace the AIM-9L/M Sidewinder in RAF service, satisfying SR(A) 1234. On 20 September 1990 the United Kingdom MoD Equipment Policy committee gave the go-ahead for the ASRAAM, allowing prime-contractor BAe Dynamics (later MBDA – Matra British Aerospace Dynamics Alenia) to launch the development program for the missile. It was planned for SRAAM to be the NATO Sidewinder replacement and to be procured by the US (United States) as the AIM-132. Following the German, Norwegian and Canadian withdrawal from the program and the US decision to develop the Sidewinder further, in November 1990, the MoD invited bids from contractors to supply a Sidewinder replacement with bids to be submitted by 6 August 1992.

As well as BAe Dynamics/Hughes (later MBDA/Raytheon) with the ASRAAM, the GEC Marconi/MATRA MICA IR (Infrared) guided air to air missile was put forward, as were designs from Loral and Raytheon in the US and Germany's BGT. At least two of these proposals were based on upgrades of the Sidewinder. On 3 March 1992, the British defence minister announced that ASRAAM had won the competition and BAe Dynamics had been awarded a contract for production of the missile.

Previous page top: the advanced IR seeker-head allows ASRAAM to successfully discriminate between the target and countermeasure decoys. BAE Systems
Previous page bottom: An ASRAAM drill round is used for training with the Leuchars Tornado F.3 Wing. Author

This page: The first operational platform for the ASRAAM was the RAF Panavia Tornado F.3 fighters, which were upgraded through the CSP (Capability Sustainment Program). This Tornado F.3 is seen with four ASRAAM on the inboard wing station stub pylons in 2002. G H Lee

At the cutting edge of SRAAM (short-range Air to Air Missile) technology, ASRAAM is an extremely agile wingless low drag missile and uses a combination of body lift and rear aerodynamic control to achieve its high performance and agility. Controlled by a software-driven autopilot, ASRAAM maintains its high agility throughout the missile flight, unlike Thrust Vector Controlled missiles. Propulsion is provided by a Royal Ordnance smokeless (low signature) solid fuelled rocket motor, which accelerates the missile to very high speed, claimed as being "hypersonic" and the fastest speed in its category.

For control the missile employs four small control fins at the rear of the body producing less drag than other SRAAMs currently in service or under development such as the European IRIS-T,

During the ASRAAM development program, the missile was integrated onto the Lockheed Martin F-16 Fighting Falcon with initial test launches conducted from F-16's in the USA. This F-16 is carrying four ASRAAM's – two each on the outer wing and wingtip stations. MBDA

Israeli Python 5 or the US AIM-9X. The low drag coefficient gives ASRAAM a longer engagement range than other contemporary systems. While missiles in the Rafael Python 4/5 class offer near instantaneous manoeuvrability off the launch rail, these weapons are short legged compared to ASRAAM and the longer motor burn of ASRAAM reduces its turning circle for close-range high off-boresight engagements. A Thorn EMI active IR, laser proximity and impact fuse actuates the DASA (now EADS Germany) developed warhead. This increases the missile effectiveness against small targets such as cruise missiles.

In modern within visual range air combat the ability to strike first is vital. ASRAAM provides an advanced capability to defeat potential adversaries, including high speed, reaction time and agility to ensure the maximum kill potential, with survivability. From close-in combat ranges to well beyond visual range, ASRAAM provides all-round target designation with full acquisition anywhere in the forward hemisphere, and the option of lock-on-after-launch engagement in both forward and rear hemispheres.

The first British Eurofighter development aircraft, DA2, with dummy ASRAAM's in the early 1990's. BAE Systems

Previous page top: ASRAAM was to fully replace the AIM-9L as the standard short-range air to air missile arming the RAF Boeing/BAE Systems Harrier GR.7/9/T.10/12 STOVL (Short-Take-Off and Vertical Landing) ground attack fighters. The Harrier, however, was retired from RAF service in 2011. Previous page bottom: A Eurofighter Typhoon from No.3 Squadron RAF launches an ASRAAM against a flare pack towed by a Mirach target drone in March 2007. MBDA Above: A RAF No.17 (Reserve) squadron Typhoon manoeuvres during a training flight revealing a pair of ASRAAM's carried on the outer wing stations. RAF

Long and short-range target acquisition and track are achieved due to the Raytheon developed wide off-boresight gimballed 128 x 128-element focal plane array IIR (Imaging Infrared) seeker and state of the art image processor. The seeker provides real-world imagery, extended acquisition range and unprecedented countermeasure resistance.

For maximum operational flexibility ASRAAM can be cued to the target using the launch aircraft's radar, a Helmet Mounted Sight (HMS) or by Data-Link at off-boresight angles of up to 90 degrees. However, the Tornado F.3 CSP aircraft were not equipped with HMS systems; therefore, ASRAAM's high off-boresight potential could not be fully exploited by the Tornado launch aircraft.

ASRAAM can also operate in Scan Mode, providing the pilot with an autonomous Pseudo IRST (Infrared Search and Track system). This facility increased the Tornado F.3's capability as the aircraft previously lacked an IRST capability.

The missiles high speed, combined with unique all round targeting capability can permit destruction of the target aircraft before it can launch its own missiles. The lock-on after launch capability allows ASRAAM to engage targets approaching from behind and the missile can regain target lock-on from the last known track even if the targets IR signature has been lost.

A RAF No.29 (Reserve) squadron Eurofighter Typhoon armed with an ASRAAM on the port outer wing station. Author

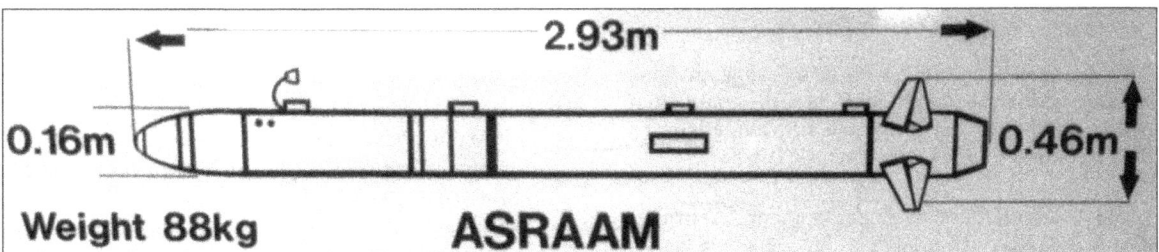

2.93m

0.16m

0.46m

Weight 88kg   ASRAAM

Top: Australia was the first export customer for the ASRAAM, which was purchased to arm the RAAF Boeing F/A-18A/B Hornet strike fighters, partnering the AIM-120 AMRAAM medium-range air-to-air missile. The Australian ASRAAM purchase gave the RAAF the most advanced short-range air-to-air missile capability of any air force in the Asia/Pacific region. This Australian F/A-18B is carrying six ASRAAM's – two under each wing on double launch rails and one on each wingtip launch station. RAAF Above: This diagram shows the basic dimensions and mass of the ASRAAM. Author

ASRAAM was initially scheduled to enter service with RAF Tornado F.3 interceptor squadrons from 1998. However, the program suffered a number of delays and in summer 1999 the prime-contractor offered the UK MoD an improved processor, which would enhance the missiles capability. The seeker head would be given a better field of view and the missile's maneuverability would be enhanced.

By early 2002 the first ASRAAM rounds had been received by the RAF and Tornado F.3 squadrons began training with the new missile in April that year. Initially it was planned that the RAF Harrier GR.7/9 squadrons would also to be equipped with ASRAAM and the missile-entered service with RAAF (Royal Australian Air Force) upgraded Boeing F/A-18A/B Hornet strike fighters in 2004. The first RAF Eurofighter Typhoons delivered in 2003 begun flying with ASRAAM training rounds as part of the build up to operational capability of the Tranche 1 Typhoon T MK1 and arms later Typhoon T.3 and FGR.4's.

The ASRAAM development offered unsuccessfully to meet the US AIM-9X requirement featured thrust vectoring to increase its manoeuvrability in high off-boresight engagements. This could feature in possible future variants of ASRAAM.

MBDA ASRAAM

Propulsion: solid rocket motor
Length: 2.90-m
Diameter: 0.166-m
Weight: 88-kg
Warhead: fragmentation explosive
Fuses: laser proximity and impact
Homing head: Imaging infrared 128 x 128-element focal plane array
Speed: claimed as hypersonic

MBDA R.550 Magic 2

Although France has developed an infrared guided variant of the MICA the MBDA R.550 Magic 2 short-range IR guided air to air missile will remain in service for a number of years, forming part of the extensive weapons options for the Dassault Rafale and Dassault Mirage 2000-5/-5MK2 (-9) multi-role strike fighters and continues to serve on other platforms including the Dassault Mirage F.1 and Dassault Super Etandard *Modernise* strike fighters.

In its original form the Magic entered service in 1975 as the only real competitor to the AIM-9 in the west. The weapon achieved considerable export success, being integrated onto 19 different aircraft types in 19 countries. The Magic also formed the basis for the Chinese PL-7 and South African U-Darter short-range IR guided air-to-air missiles.

The current Magic 2, which has been in operational service since 1985, is a second-generation Magic featuring dramatically increased performance over the first generation missile. The SAT AD.3601 passive scanning IR seeker of the

This Dassault Mirage 2000-5 multi-role strike fighter is armed with four under fuselage MICA EM radar guided air-to-air missiles and a pair of Magic 2 IR guided short-range air-to-air missiles on the outer wing stations. Dassault

Magic 1 was replaced by the AD.550 multi-cell unit, cooled from a nitrogen tank in the launch rail. The seeker has a +/- 70-degree off-boresight launch capability. The Magic 2 also introduced all-aspect interception capability with a range of 550-metres out to 20+km (10.8-nm). The increase in range was achieved by a 10% increase in rocket-thrust. A digital autopilot with a built-in microprocessor controls the proportional navigation unit prior to IR (Infrared) homing in the end-phase. The pilot can select sensor integrated or autonomous acquisition modes depending on the tactical requirement during an engagement. The seeker searches autonomously in a pre-defined wide or narrow scan fields or can be slaved to designated targets.

The Magic 2 can manoeuvre at up to 55-g and together with its 12.5-kg (27.6-lb) pre-fragmented warhead, which is detonated by an RF (Radio Frequency) Doppler proximity fuse (with triggering delays according to missile target closing speed), ensures a high kill probability.

The Magic 2 has a length of 2.75-m (9-ft), a diameter of 16-cm (6.3-in), a span of 66-cm (26-in) and a launch weight of 89-kg (126-lb).

The Magic 2 was the standard infra-red guided short-range air-to-air missile used by French Air Force Mirage 2000C interceptors and Mirage 2000D strike aircraft. This two-seat Mirage 2000F, taking off from Farnborough in the United Kingdom in the 1980's, is carrying Magic II missiles son its outboard wing stations. US DoD

Previous page top: When the Dassault Rafale M entered service with the French Navy in the early 2000's the MICA IR had not been cleared for service, therefore the Magic 2 was the standard IR guided missile for the services Rafale's. Even after the introduction of the MICA IR, the Magic was retained as part of Rafale's armament options. Dassault Previous page bottom: A Magic 2 is carried on the port outer wing station of a French Navy Dassault Super Etandard *Modernise*. MBDA

This page top: The French Air Force Mirage 2000D/N strike aircraft are armed with Magic 2 missiles for self-defence. EADS Above: The Magic 2 at top of photograph serves alongside the more modern MICA IR at bottom of photograph. Author

Previous page top: A French Air Force Mirage 2000C takes-on fuel from an USAF tanker aircraft during operation Allied Force in April 1999. The Mirage is armed with Magic 2 missiles on the outboard wing stations. USAF Previous page bottom: The first production Rafale B for the French Air Force, B301, is armed with a pair of Magic 2 missiles on the wingtip launch rails. Dassault

This page: The first production Dassault Rafale M for the French Navy with Magic 2 missiles on the wingtip launch rails. The Magic 2 and the more modern MICA share common fixtures allowing them to use the same launch rails. Dassault

On the Rafale strike fighter the Magic 2, like the MICA IR, can be integrated with the Sextant Avionique third-generation binocular Topsight E HMSS (Helmet Mounted Sighting System), which commenced development in the mid-1980's. The Topsight HMSS combined target detection and image generation in a single unit. The system has 360-degree azimuth coverage from a cockpit-magnetic tracking system. The visor displays flight and sensor data as well as target tracking and acquisition. The system has a reported weight of 1.54-kg (3.2-lb).

Magic 2 is used by a number of air forces including Greece, Egypt, Spain, Kuwait, United Arab Emirates and the French Air Force and Navy. The missile is integrated with Dassault Mirage F.1, Mirage 2000B/C/D/N, Mirage 2000-5/-5MK2; Dassault Super Etandard and Dassault Rafale strike fighters.

MBDA R.550 Magic 2

Length: 2.75-m
Diameter: 0.16-m
Weight: 89-kg
Speed: Mach 2.7
Range: around 8 miles
Warhead: 12.5-kg high explosive blast fragmentation
Fuse: RF (Radio Frequency) proximity
Guidance: infrared seeker with all-aspect engagement capability

The MICA EM active-radar guided variant completed development ahead of the MICA IR variant and became operational on French Mirage 2000-5F and export variants of the Mirage 2000-5/-5MK2 in the late 1990's. A MICA EM is housed on the port shoulder station of a French Air Force Mirage 2000-5F in 1997. Author

MBDA MICA EM/IR

For air-to-air engagements the Dassault Rafale and Dassault Mirage 2000-5/-5MK2 (2000-9) employ the MATRA BAe Dynamics MICA (*Missile d' Interception de Combat et d' Autodefence* - Dogfight/Self-Defence/Interception Missile) air-to-air missile, which is available in two variants; passive imaging infrared (heat seeking) or active radar guided EM. Both variants share a common airframe, warhead and propulsion system, but are equipped with different homing systems and countermeasures systems. The weapon was described as the first multi-target/multi-mission air-to-air missile system capable of intercept missions beyond visual range. The Dassault Rafale fighter can carry an operational load of six of these

formidable weapons as can the Dassault Mirage 2000-5/-5MK2.

MICA was designed as replacements for the MATRA Super 530D medium-range semi-active radar guided and Magic 2 short-range infrared guided missiles employed on the Mirage 2000C. The Magic 2, however, remained in the inventory as a part of the Rafale's armoury, supplementing the MICA.

Above: The MICA airframe was used for a modular family of air-to-air missiles with active-radar homing and IR homing heads. The MICA EM is nearest camera with the MICA IR behind. Author

The Mirage 2000-9/-5MK2, above, can carry a total of 6 MICA's, with a typical combat load consisting of four MICA EM's on the shoulder stations and a pair of MICA IR's on the outboard wing stations. MBDA

The MICA IR has a SAT/MATRA passive IIR homing head sensor and the MICA EM has a Thales (formerly Dassault Electronique) AD4A active radar seeker head operating in the 'J' band, suitable for medium range engagements. Once the launch aircraft's onboard radar has designated a target the missile is launched on the inertial guided first phase of the flight; the missile then locks onto the target using its onboard active radar homing-head. MICA's multi-target fully autonomous fire and forget capability allows the pilot to engage several targets simultaneously even in a dense electronic warfare environment and saturation jamming.

The missile has a launch weight of 112-kg (244-lb), length of 3.1-m, a span of 56-cm, a diameter of 16-cm and a range of 50-60 km. Speed is classified, but is thought to be around Mach 4 with excellent acceleration provided by the solid rocket engine. The keep the missile to the desired launch weight the high explosive blast fragmentation warhead is considerably less powerful than rival

weapons such as the US Raytheon AIM-120A AMRAAM.

The MICA missile has excellent manoeuvrability, which had been demonstrated in over 100 test-launches by the early 2000's. The missile uses a combination of aerodynamic control surfaces and thrust vector control to achieve it's exceptionally agility with manoeuvre load limits of up to 50-g.

A MICA IR fitted to the port wingtip station of a French Naval Dassault Rafale Omni-role fighter aboard the aircraft carrier *Charles de Gaulle*. MBDA

Previous page top: A Dassault Rafale M banks to port revealing its maximum load of six-MICA's (mixed variants). MBDA Previous page bottom: As part of the MICA development program a large number of missiles were launched from Rafale development aircraft. A Rafale M launches a MICA missile from a starboard wing station during a development firing. EADS

This page above: A Rafale launches a MICA missile in the late 1990's. MICA was the first European designed active-radar guided air-to-air missile to enter service. EADS

When launched on medium or long-range engagements the active-radar guided MICA EM uses strap-down inertial guidance, which can be updated via a data-link from the launch aircraft if required during long-range engagements. The missiles onboard active radar seeker head is only activated during the extreme terminal phase of the engagement. For short-range engagements the missile is locked onto the target aircraft prior to launching.

The MICA IR variant is the longest-range weapon of its type in the western world and incorporates excellent angular resolution (duel band imagery) and stealth. When used with the Dassault Rafale's FSO (Front Sector Optronics) IRST system the MICA IR's passive homing head enables absolutely "silent" interceptions to be conducted. This bestows a great advantage for the launch aircraft by attacking the enemy before they are aware of the Rafale's presence. The MICA IR can also be used by the pilot for passive IR monitoring complementing the active monitoring of the RBE2 radar during the mission.

The heart of MICA EM's capabilities is the AD4A active-radar seeker, which is also used as the basis of the ASTER Surface to Air Missile and Meteor extended-range beyond visual range air-to-air missile seeker heads. MBDA

Above: A Dassault Rafale armed with 2 MICA IR, 4 MICA EM and an Exocet anti-ship missile. Below right: A Mirage 2000 from the French DGA is carrying four MICA EM radar guided missiles on the fuselage/wing root stations. DGA

The MICA EM can be used in four different engagement modes:

- Long-range multi-target interception
- Multi-target interception at medium-range
- Short-range air combat
- Self-defence of aircraft on tactical missions

MATRA began development of the MICA in 1975 and the weapon was from the beginning developed as both a BVR and close range dogfight missile able to replace both the R.530 and R.550 missiles then in use. Although the weapon is not as manoeuvrable and weighs in heavier than dedicated short-range missiles, the MICA weighs considerably less than other BVR weapons. With a launch weight of 110-kg the MICA is 40-kg lighter than the Raytheon AIM-120, which weighs in at 150-kg. While the MBDA ASRAAM, with a launch weight of 85-kg, is considerably lighter than MICA, the latter has the advantage of a longer range.

Matra (now MBDA) received a development contract for MICA in April 1987 and the weapon began flight-testing in 1994. In 1995 Matra carried out its first air launch of a MICA IR missile equipped with a SOFRADIR imaging infrared (IIR) guidance system. The missile, which was launched from a Mirage 2000, scored a direct hit against an Aerospatiale CT.22 target drone, without the use of data-link target update information. MATRA claimed that the range this was achieved at was greater than any IR guided air-to-air missile in service anywhere in the Western World, although it declined to reveal the distance involved. Basic development of the IR variant of MICA was completed in 1997.

The MICA IR, here carried on the starboard wingtip station of a Dassault Rafale, bestows upon that fighter the longest-range passive IR interception capability of any current western combat aircraft. Dassault

Matra BAe Dynamics received orders for more than 2000 MICA missiles from France, Qatar and Taiwan for use on the Mirage 2000-5 and 2000-5 MK2 (2000-9) and further orders are were received from the French government to equip the Mirage 2000-5F and Rafale. The MICA EM first entered service on the Mirage 2000-5F serving with 1/2 'Cigognes' and 2/2 'Cote d'Ore' of the French Air Force at Dijon in France.

Following a series of 27-succesful test firings of MICA from a Rafale at the CEV (Flight Test Centre), full integration of the missile on the Rafale F1 Standard fighter was completed in July 2000. MICA EM's were the primary armament of the first F1 Standard Rafale M's to enter service with the Aeronavale revolutionising the services air defence capability as the service previously lacked even a SARH (Semi-Active Radar Homing) air to air missile capability. The introduction of the Northrop Grumman E-2F Hawkeye AEW (Airborne Early Warning) aircraft for service from the French aircraft carrier *Charles de Gualle* has further enhanced the air defence and air superiority capability of the Aeronavale. The Rafale M F1 Standard fighter combined with its MICA EM

missiles bestowed upon the French Navy, without a doubt, the most capable air defence capability in any naval air arm in the world when it entered service in 2001.

Both radar and IR variants of the MICA can also be integrated on the Mirage F.1 fighter as part of upgrade packages.

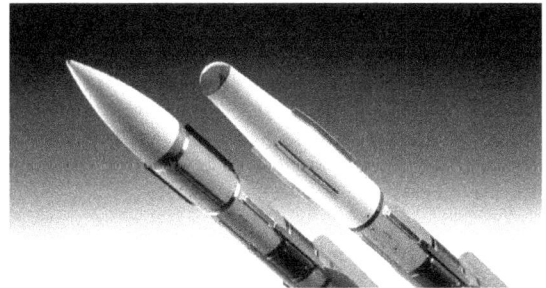

The MICA program was aimed at fielding a modular family of missiles featuring both IR and active radar homing. The MICA IR and EM are identical apart from the different seeker heads. MBDA

MBDA MICA

Length: 3.10 metres
Diameter: 0.16 metres
Weight: 112-kg
Homing heads (x2): Active EM or passive imaging infrared
Range: around 60-km

The Meteor extended-range BVR active-radar guided air to air missile emerged from a RAF requirement for a mew missile to arm the Eurofighter Typhoon from around 2012. The program has grown into a pan-European missile program, with **MBDA** as prime-contractor, to field the weapon for the airforces of France, Germany, Italy, Spain and Sweden as well as the UK. This artist impression shows a Meteor missile being launched from the starboard rear shoulder recess of a RAF single-seat Eurofighter Typhoon. MBDA

MBDA Meteor

In the late 1990's, the main driving force behind European and US BVRAAM (BVR Air to Air Missile) development work were primarily aimed at winning the UK contract for SR(A)1239 (Staff Requirement Air 1239) for a BVRAAM to arm the Eurofighter Typhoon. In late 1995 a RFP (Request For Proposals) was issued by the UK MoD for SR(A)1239 to meet the RAF requirement for a FMRAAM (Future Medium Range Air to Air Missile). In the mid-1990's RFI (Requests For Information) had brought responses from a European consortium led by BAe Dynamics with a

development of the its S225X, Matra of France, with a development of the MICA EM and Daimler-Benz Aerospace Bayern Chemie with the A3M.

In its original form, a wingless configuration was chosen for Meteor for optimum aerodynamic and kinetic performance with an electrically actuated tail control system to provide a high degree of agility and minimum drag. An early wingless Meteor model is shown on a Saab Gripen mock-up in 1996. Author

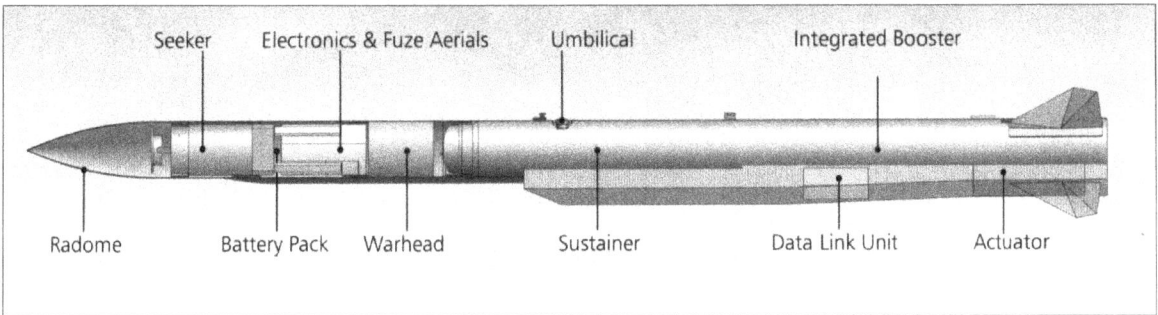

Top: This early wingless configuration of the Meteor design was based on studies conducted in the S225X program. Author

Above: This cutaway diagram of the wingless Meteor design dates back to 1996. However, the internal systems layout of the missile has changed little. MBDΛ

The FMRAAM (Future Medium Range Air to Air Missile) requirement called for a missile with a range of around 80-nm. This resulted in most of the proposals employing rocket ramjet propulsion; although BAe Dynamic's studies of its S225X program in conjunction with GEC Marconi, Alenia and Saab Missiles, encompassed a range of rocket, ramjet and compound rocket/ramjet propulsion options. The S225X set out to provide fire and forget capability from data-link, mid-course guidance and active radar homing, plus high ECM (Electronic Countermeasures) resistance to meet agile evasion tactics and stealth characteristics. S225X options developed within the size constraints of the Eurofighter Typhoon conformal fuselage bays were quoted as including a duel-

section boost-coast-boast cycle rocket motor to provide it with a high average velocity and optimum terminal speed. A wingless configuration was also specified for optimum aerodynamic and kinetic performance, with an electrically actuated tail control system to provide a high degree of agility and minimum drag. Alenia field-tested active pulse-Doppler radar with a wide-band gallium arsenside transmitter for its agile solid-state seeker (AS3), in parallel with research by the other S225X partners, including GEC Marconi with active radar seeker work. GEC Marconi drew on active radar technology it had developed for the French MATRA MICA EM AD4A radar seeker.

In early 1996, a European Consortium led by BAe Dynamics, which included Saab, GEC Marconi and Alenia, which were previously teamed to offer the S225X, joined with MATRA, Daimler Benz, Aerospatiale and LFK to offer the Meteor missile for the RAF's FMRAAM requirement. The Meteor program was officially launched at Farnborough in September 1996.

Previous page top: Towards the end of the 1990's the Meteor design team began to incorporate two short-span mid-set wings into the design. MBDA **Previous page bottom: One of the major design drivers for the Meteor team was to design a weapon within the constraints of the Eurofighter Typhoon fuselage shoulder missile bays. A Meteor mock-up is shown on a mock-up of a German Typhoon in 2001.** Author

This page above: The Saab Gripen can carry Meteor missiles on up to four wing stations used for the carriage of AIM-120 AMRAAM. A wingless configuration Meteor is shown on a portside wing station of a Saab Gripen. Saab

Two teams competed for SR(A) 1239. Following a series of mergers and acquisitions the European Meteor consortium of Matra BAe Dynamics, DASA's LFK division, Alenia Difesa, CASA, Marconi Electronics and Saab combined to offer the Meteor. Raytheon offered a phased program of AIM-120B+ AMRAAM, ERAAM+ (Extended Range Air to Air Missile Plus) and then FMRAAM (Future Medium Range Air to Air Missile). Matra BAe Dynamics claimed that Meteor would be three times more effective than then current MRAAMs (Medium Range Air to Air Missiles) like AMRAAM, offering higher performance than that required for SR(A) 1239.

Both the European Meteor and the Raytheon FMRAAM were initially externally similar with the main difference being internal, particularly with the propulsion unit. While the propulsion selected for Meteor was a solid-fuel variable flow ducted rocket-ramjet, the Raytheon FMRAAM would have employed a rocket, which after burnout would switch over to a ramjet-sustainer.

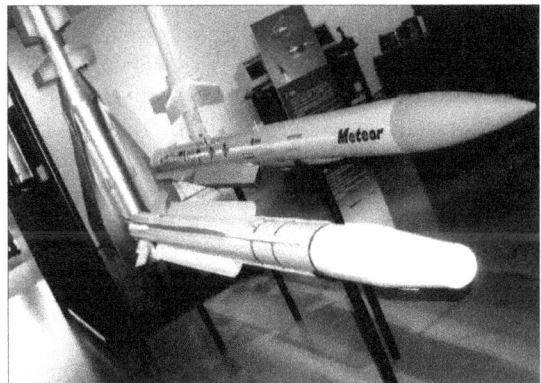

Meteor is being designed for easy integration onto aircraft currently operating with other air-to-air missile systems such as the MBDA MICA. On the Dassault Rafale the long-range of Meteor will be complemented by the long-range capability of the MICA IR seen above with a Meteor. Author

Above and right: When it enters full operational service Meteor promises to be the highest performance air-to-air missile in the world, featuring enhanced capabilities including a vastly increased 'no escape zone' over missiles like AMRAAM. These early mock-ups date back to the late 1990's; the missile featuring centre body fins. Author

The rear section of the Meteor comprises the Bayern Chemie/Protac Boron-filled, solid propellant, and throttle, ducted ramjet motor. High speed is maintained throughout the flight and the motor offers variable thrust levels to match the hard manoeuvres of the target. Meteor is capable of autonomous engagement of targets ranging from fighters to large transport size aircraft down to small cruise missiles, in all-weathers day or night, using its active seeker, even in a dense electronic countermeasure environment. The ramjet propulsion ensures the missile has a range thought to be in excess of 100-km and a speed in excess of Mach 4; therefore, even when launched at maximum-range, the missile will retain enough energy in the 'end game' to defeat fast manoeuvrable targets such as fourth and fifth generation fighters and advanced missiles conducting evasive manoeuvres.

The GEC Marconi (now part of BAE Systems) AD4A active-radar seeker was selected as the basis for developing the terminal guidance seeker and in June 2003 an agreement was signed by Thales and MBDA to jointly develop an improved variant AD4A seeker, also used in the MICA EM air to air and Aster SAM (Surface to Air Missile) programs.

This artist impression shows a Dassault Rafale launching an out of date iteration of the Meteor from a fuselage shoulder station. The aircraft is shown carrying MICA IR missiles on the wingtip stations. Meteor will be operated alongside the MICA IR and EM. MBDA

Meteor's powerful warhead increases the probability of achieving a kill against airborne targets, which would be destroyed by the blast fragmentation warhead, which is activated by proximity or contact fuzzes.

The sale pitch was that Eurofighter Typhoon equipped with Meteor is Europe's cost-effective solution to defeating current/future threats at a fraction of the cost of the US response - the Lockheed Martin Boeing F-22A Raptor armed with the AIM-120 AMRAAM. Meteor offers a 100% European solution, providing an opportunity to develop and exploit a world-class European missile industry, creating choice and competition. Computer simulations show that when Meteor is fitted to fighters in the Eurofighter class, Eurofighter defeated the predicted Russian/Chinese threat every time - an unbeatable combination, which will defeat all threats for the foreseeable

future at a fraction of the cost of the F-22A. However, it should be stressed that these simulations are based on currently available information and not on much of the future fighter and missile technology that is being developed in Russia and China. What is clear is that Meteor is far more capable than the AIM-120 AMRAAM or the Russian R-77M (AA-12 'Adder') with a significantly improved 'no escape zone'.

This retouched photograph shows a Rafale carrying four Meteor missiles. Dassault

This retouched photograph shows a Meteor being launched from a RAF Eurofighter Typhoon. MBDA

Development of a new missile in the Meteor class was essential to keep the Dassault Rafale, Eurofighter Typhoon and Saab Gripen ahead of the competition. Russian missile technology is very advanced, and high performance weapons like the Vympel RVV-AE (known in the west as the R-77 AA-12 'Adder') have been delivered to export customers. These include India, Malaysia, Peru and China is also reported to have purchased the weapon for its Sukhoi Su-30MKK, MK2/3 and Su-27 'Flanker' fighters. The R-77 has also been in limited service in Russia, possibly since 1994.

The UK BVRAAM competition was only the first of a number of European competitions for similar weapons. The remaining Eurofighter partner nations, Germany, Italy and Spain as well as France and Sweden all required a BVRAAM for Eurofighter Typhoon, Rafale and Saab Gripen fighters respectively. The European consortium competing for the UK contract pointed out that Meteor was essential to Europe retaining control over export of its combat aircraft, which if armed with US missiles would be subject to US Congressional Approval for export sales. This argument won through when the UK MoD selected the Meteor in June 2000, with MBDA chosen as prime-contractor for the European program.

This air defence configured Dassault Rafale is carrying Meteor BVRAAM on the intermediate wing stations and MICA IR missiles on the wingtip stations. The aircraft is also carrying three external fuel tanks, used for long-rang ferry flights or to extend its operational combat radius. MBDA

Selected by the UK, Meteor was offered to satisfy a number of European competitions for similar weapons to arm their Profiteer, Rafale and Gripen. As noted, the European consortium competing for the UK contract had previously pointed out that Meteor was essential to Europe retaining control over export of its combat aircraft, which if armed with US missiles would be subject to US Congressional Approval for export sales. This argument, combined with Meteors and the missile and France, Sweden and the United Kingdom gave the official go-ahead to the Meteor program at the Paris air show in June 2001. At the same time, Germany, Italy and Spain confirmed an intention to sign up to the program with the Italian government approving funding for Italy's participation in August 2001.

Italy planed to purchase 400 Meteor's with a 12% share of development work; Germany had 21% of development costs, while the UK invested around $1.5 billion in the Meteor program.

In late 2002, the UK concluded an agreement to take an additional 5% share of Meteor development costs to offset Germany's reduction following a budget meeting on 18 December that year. This meeting confirmed a revised number of 600 Meteor missiles for Germany, down from the originally planned 1,488. This reduction reduced Germany's share in the program to 16%. The UK share was increased to 39.6%, with France 12.4%, Italy 12%, Spain 10% and Sweden 10% remaining unchanged. The Meteor fixed price development contract was signed by all six-partner nations on 23 December 2002.

Meteor development was planned to take around eight years following which the missile WAS expected to enter service on the Eurofighter Typhoon, Saab Gripen and Dassault Rafale from around 2012. Throughout its gestation period the Meteor has undergone many changes, such as the introduction of small fins to the missile body, before the definitive missile emerged.

Functional integration is not expected to cause any major problems for the Gripen and Typhoon currently armed with AMRAAM or the Rafale armed with MICA, as MICA AMRAAM and Meteor all operate in the same way. The Rafale data-link, optimised for MICA was fully capable of accepting the Meteor.

Early Meteor/Eurofighter Typhoon integration work included fit trials of the weapon conducted on the British built Eurofighter development aircraft DA4 at Warton. MBDA

The Rafale is capable of carrying up to four Meteor missiles, which would normally be carried on the intermediate wing stores stations and on the lateral fuselage stations. The Rafale will also continue to operate with MICA in both EM and IR variants. The Gripen will also be capable of carrying four Meteors, while the Typhoon will be capable of carrying up to eight.

In 2004 a Meteor was successfully fitted to a Saab Gripen during a series of trial fits, and the second series of aerodynamic and wind-tunnel tests were conducted that year. The following year a series of live catapult take-offs and landings were conducted aboard the aircraft carrier *Charles de Gaulle,* with a Dassault Rafale fitted with Meteor mock-ups in order to confirm the missiles handling characteristics.

Meteor Test firings commenced in 2006 when a series of 'lower altitude' Meteor firings were conducted from a Saab JAS 39 Gripen, followed by completion of 'seeker data gathering trials in early 2007. A high altitude Control and Dispersion (C&D) test launch of a Meteor was conducted at

QinetiQ's MoD firing range in the Outer Hebrides on 22 May 2007. During this launch a Meteor round was 'rail-launched' from a Saab Gripen, when the aircraft was flying at supersonic speed at an altitude of 42,650 ft (13 km). Following a few seconds of boost, the missiles VFDR (Variable Flow Ducted Rocket) intakes were opened allowing the missile to transition to the ramjet, whereby it then accelerated to speeds in excess of Mach 3, while conducting "various challenging manoeuvres including using the novel bank while turn in control algorithms," which were developed specifically for Meteor. During the flight the Meteor "followed a pre-programmed flight profile for several minutes, which also demonstrated the high end game control capability for missile launches even at the maximum kinetic range." Following the successful flight, the meteor was then deliberately "broken up in flight" over the firing range.

In early July 2012, MBDA concluded the Meteor guided firing trials with three successful launches, each achieving direct hits against targets deploying countermeasures. The trials were conducted under EPM (Electronic Protection Measure) trials aimed at proving the missiles capability against targets in an operational environment.

Top: This Swedish Saab JAS 39 Gripen, 101, was extensively used during the Meteor development test firing program. Above: A mock-up of the Lockheed Martin F-35 Lightning II is shown with a Meteor missile. The F-35 will carry the weapon on external wing stations. MBDA

Top: Italian Typhoon Instrumented Production Aircraft IAP2 carrying AIM-9L and a mock-up Meteor missile on the starboard intermediate wing station. Eurofighter GmbH Above: A Saab Gripen conducts a Meteor missile launch around 2007. Right: a Meteor development round is fired from a QinetiQ operated Panavia Tornado F.3. MBDA

In total, 21 air launched firings of Meteor missiles had been completed by this date; divided into two main phases: Development firings 2006 – 2008 and Guided firings 2009 – 2012. The development firings were conducted from a Saab JAS 39 Gripen, while the guided firings were conducted by a Gripen and a Panavia Tornado F.3 operated by QinetiQ.

Supporting the Meteor firing trials, in excess of 40 flights were conducted under the seeker data gathering phase of the program. Over 100 "ground firings of the throttleable ducted rocket propulsion system", were also conducted.

The first launch of a Meteor from a Eurofighter Typhoon was conducted in early December 2012 under the Future Enhancements Flight Test Program. The Meteor was launched from one of the Typhoons rear semi-conformal fuselage missile stations.

In May 2013, Germany signed a production contract ordering the Meteor, joining the UK, Italy, Spain, France and Sweden. Production deliveries were expected to commence later in 2013.

Meteor

Range: in excess of 100-km
Speed: in excess of Mach 4
Length: 3.67-m
Propulsion: Bayern Chemie variable flow ducted rocket ramjet
Aerodynamics: cylindrical with an asymmetric ramjet air-intake configuration. Two mid-mounted wings and 4 aft mounted fins.
Seeker: MBDA/Thales developed active radar
Navigation/guidance: inertial mid-course with data linking. Autonomous terminal guidance using advanced proportional navigation
Warhead: blast fragmentation
Fuses: Saab Bofors Dynamic's proximity fuse and impact fuse

A MBDA Skyflash SARH medium range air-to-air missile is launched from a Panavia Tornado F.3 in the late 1980's. Skyflash is augmented by the Raytheon AIM-120 AMRAAM. BAE Systems

MBDA Skyflash

When it entered service the RAF Panavia Tornado F.2/3 Interceptors primary armament consisted of four BAe Dynamics (now Matra BAe Dynamics Alenia) Skyflash SARH (Semi-Active Radar Homing) air-to-air missiles housed under the fuselage. Skyflash had replaced the earlier US Hughes (now Raytheon) AIM7E-2 Sparrow SARH BVR on the RAF Phantom FGR.2/FG.1 fighters before being carried over to the Tornado F.2/3. Skyflash was based on the body of the Sparrow air to air missile, but with an entirely new British motor and seeker head.

Skyflash missiles were delivered to the RSAF (Royal Saudi Air Force) to arm that services Panavia Tornado F.3 interceptors purchased from the UK. Sweden also ordered Skyflash missiles as the primary armament for its JA 37 Viggen fighters. In Swedish service the Skyflash was

designated Rb71 and it was expected that the Rb71A Active Skyflash (Skyflash 90) would be ordered before Sweden switched its attention to the US AIM-120 AMRAAM. The Italian Air Force took delivery of a small stock to arm Tornado F.3's leased from the RAF in the mid-1990's.

Skyflash missiles are carried on four fuselage launch stations on the Tornado F.3 as seen in this view of a Tornado F.3 on QRA Duty at RAF Leuchars in 1993. Author

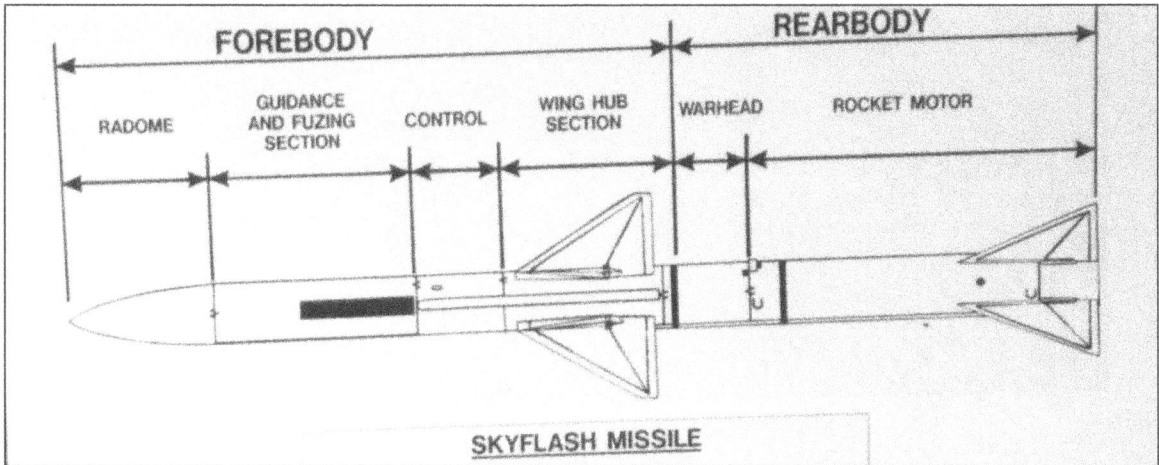

FOREBODY   REARBODY

RADOME | GUIDANCE AND FUZING SECTION | CONTROL | WING HUB SECTION | WARHEAD | ROCKET MOTOR

SKYFLASH MISSILE

Previous page top: A RAF No.43(F) Squadron Panavia Tornado F.3 reveals its underside showing the staggered formation carriage of the primary armament of four Skyflash missiles. The forward pair of missiles is semi-recessed. Crown Copyright **Previous page bottom: A Skyflash and AIM-9L Sidewinder drill round sit in front of a No.111(F) Squadron Tornado F.3 at Leuchars in 2002.** Author

Above: Diagram showing the layout of a Skyflash missile, which is made up of two main sections divided into five sub-sections between them. At the front is the radome followed by the guidance and fusing section, control section, wing hub section, warhead and the propulsion section. Author

Skyflash was designed as a supersonic medium range air-to-air missile. The latest variants of the missile incorporated a boost-sustain solid fuel rocket motor increasing range over earlier variants. The missile can be employed against targets in all-weather conditions and has the ability to snap-up or snap-down allowing engagement of targets at 'ultra-high' or 'ultra-low' levels. The seeker-head can discriminate between separate target groups and can operate in a dense counter measure environment.

The Tornado F.2/3 can carry four Skyflash missiles on the under-fuselage stations, with the front pair of missiles being semi-recessed. When launched, the missiles are pushed downward into the airflow by large Frazer Nash rams. Once launched the missiles use semi-active radar homing, whereby it heads towards the target area guided by the Tornado's AI.24 Foxhunter radar. The launch aircraft has to continually illuminate the target allowing the Skyflash to home in on the targets radar returns. Once near the target Skyflash detonates using a proximity fuse. Although primarily used as a BVR weapon Skyflash can also

be used against targets at shorter ranges. In this type of engagement, the missile is optimised for quick reaction and maximum manoeuvrability once launched from the mother aircraft.

In February 1987, the Tornado F.3 conducted missile firings at the Aberporth ranges for the first time, achieving good results in Skyflash firings against target drones. The missile has been deployed operationally with Tornado F3's from the RAF, RSAF and Italian Air Force in the Persian Gulf and the Balkans, but none were ever fired in anger.

Above right: This No.5 Squadron RAF Tornado F.3 is armed with four Skyflash and 4 AIM-9L Sidewinder missiles. Crown Copyright **Above: A Live Skyflash is mounted on the starboard forward fuselage recess of a Tornado F.3.** Author

A No.11 Squadron RAF Tornado F.3 reveals its underside showing its four Skyflash and 2 AIM-9L Sidewinder short-range IR guided air to air missiles. G H Lee

One of the major drawbacks of semi-active radar guided air-to-air missiles is the requirement for the launch aircraft to continually illuminate the target with its radar for the entire flight of the missile. This leaves the launch aircraft vulnerable to enemy counters including air launched BVR and short range missiles as the distance between the target and the launch aircraft closes at high speeds sometimes bringing the launch aircraft within range of short-range infrared guided air to air missiles. To overcome this vulnerability the RAF had long harboured plans for introduction an active beyond visual range air-to-air missile to replace Skyflash on the Tornado F.3. This was initially planned to be an active radar guided variant of Skyflash itself, most commonly referred to as 'Active Skyflash'. However, this program was never brought to fruition and RAF interest switched to the US Raytheon AIM-120 AMRAAM active radar guided beyond visual range air-to-air missile, which was entering service on US fighters in the early 1990's.

Although effectively replaced on the Tornado F.3 by the AIM-120 AMRAAM, Skyflash stocks remained in use until the Tornado F.3 was retired from RAF service when the last operational squadron, No.11, was disbanded at RAF Leuchars on 22 March 2011. The weapon remained in service with RSAF Tornado F.3's, but will be retired when these aircraft are fully replaced by Eurofighter Typhoons in RSAF service.

MBDA Skyflash

Length: 3.66-metre
Diameter: 20.3-cm
Launch weight: 208-kg (RAF documentation also quotes a weight of 224-kg).
Range: around 30+ miles
Speed: up to Mach 4
Guidance: Uses launch aircraft radar for initial guidance before switching to on-board semi-active-radar terminal guidance seeker

---

A French Air Force Dassault Mirage 2000C armed with MBDA R.550 Magic 2 IR guided air to air missiles on the outboard wing stations and MBDA Super 530D SARH medium-range air to air missiles on the inboard wing stations. USAF

MBDA Super 530D

The MBDA (formerly Matra) Super 530D semi-active radar homing medium range air to air missile was an evolutionary follow-on to the original Matra R530 SARH missile, which began equipping French Air Force Mirage III fighter squadrons from 1963. The R530 was produced in SARH and Infrared guided variants. The SARH variant in particular, like all early radar guided air-to-air missiles, had a very poor performance and kill probability. This led to the introduction of the Matra Super 530F SARH missile which was developed as the initial primary armament for the Dassault Mirage 2000B/C interceptors serving with the French Air Force when this aircraft was delivered from the early to mid-1980's.

The Super 530F was developed primarily for operations at medium to high altitudes, while the Super 530D in service today featured improved performance capability against targets flying at low-altitude, introducing the AD26 monopulse continuous-wave-Doppler SARH seeker head. Although the 530D was a major improvement over the 530F, it was still tied to the SARH principal of the launch aircraft by having to continually illuminate the target with its on-board radar during the missiles flight to the target. This left the launch aircraft vulnerable to counter-fire as it was unable

to take evasive manoeuvres without losing target lock. As well as increased targeting capability, which enhanced the probability of destroying the target, the 530D also featured enhanced ECCM resistance compared with its predecessors.

MBDA Super 530D
Length: 3.80-m (12-ft 6-in)
Diameter: 263-mm (10.35-in)
Wingspan: 0.62-m ((2-ft 0.5-in)
Launch weight: 275-kg
Warhead: 30-kg (66.6-lb) high explosive
Speed: Mach 4.5
Fuse: radar
Guidance: AD26 SARH

Bodenseewerk Garatetechnik GmbH IRIS-T

IRIS-T is a 5th generation SRAAM developed as an AIM-9 Sidewinder replacement for the German, Italian Spanish, Swedish and other air forces as their standard short-range air-to-air missile. In 1996, six NATO nations signed a MoU (Memorandum of Understanding) to develop a new fifth-generation high-agility SRAAM as a common sidewinder replacement. The new missile project was known as the Infra-Red Imaging Seeker - Tail control (IRIS-T). Germany took shares of 46%, Italy, 20%, Sweden 18%, Greece, 8%, Canada 4% and Norway 3%. Canada withdrew from the program, but its place was eventually taken by Spain. BGT, the prime contractor, received a nine-month project definition contract from the German BWB (Defence Procurement Agency) in September 1996, this being completed in April 1997. A 48-month development phase was

Top: The original IRIS-T design featured mid-set wings of almost trapezoidal design with the same taper at both leading and trailing edges seen on this model of an early IRIS-T design in 1996. The wings are also of slightly longer span than on the definitive IRIS-T. The 1996 model also featured small strakes just behind the seeker head. Author

Above: IRIS-T has been developed as a highly agile fourth generation short-range air-to-air missile with a high sensitivity IIR seeker head. The missile is now in service with a number of NATO air forces and is being integrated onto a number of modern combat aircraft including Eurofighter Typhoon, Panavia Tornado IDS and Saab JAS 39 Gripen. The IRIS-T features long-chord wings offset toward the rear of the missile body. Author

The wings of the early 2000'S IRIS-T iterations, here in mock-up form (top), were more streamlined at the leading edges of the wing than those of the production standard missiles carried by an Italian Typhoon during operations over Libya in 2011 (above). Author and Eurofighter GmbH

launched in late 1997 and agreement for full-scale engineering development was signed on 27 April 1998.

Thrust-vector control provides IRIS-T with high agility and the missile features +/-90-degree off-boresight seeker look angles, lock-on after-launch capabilities, and image processing with resistance against counter measures and built in test features. The missiles interface with aircraft launchers are compatible with existing analogue as well as more modern digital interfaces found on fourth and fifth generation combat aircraft like Eurofighter Typhoon and Saab Gripen.

The IRIS-T is replacing the AIM-9 in German Luftwaffe service on the Eurofighter Typhoon and Tornado IDS, giving the latter aircraft an excellent self defence capability against opposing fighters and as well as air and surface launched missiles. BGT

The missile seeker can be slaved to a target by the launch aircraft's on-board radar system, an IRST (InfraRed Search and Track) system or via a HMSS (Helmet Mounted Sighting System). This latter method allows the full utilisation of the +/- 90-degree off-boresight capability of the missile.

During the design phase the overall dimensions and physical characteristics of the missile were deliberately selected to correspond with those of the AIM-9L Sidewinder reducing the potential problems with integrating the missile with a large number of combat aircraft already operating with the AIM-9. The cruciform forward stabilisers, mid-body wings and rear fins produce more drag than the Sidewinder the IRIS-T is designed to replace. These same control surfaces, in conjunction with thrust vector control, confer excellent manoeuvrability with the minimum possible turn circle. The +/- 90-degree look angle high sensitivity seeker head gives IRIS-T its high off-boresight capability. According to the manufacturer IRIS-T is 25 times more resistant to IR

countermeasures than the AIM-9L Sidewinder. Target acquisition is via a two-axis roll/pitch 3-5 micron 2 x 64-staggered high resolution IIR seeker, which has a look angle of +/- 90-degree and high IRCCM (InfraRed Counter Counter Measures) resistance. This seeker was developed from the BGT Tell IIR seeker, which was unsuccessfully submitted for the AIM-9X competition. The warhead has active laser proximity and contact fuses.

The image processing algorithms employed by the IRIS-T IIR seeker were successfully flight tested during launches of AIM-9M Sidewinder airframes fitted with the new seeker head. During 1996, a Luftwaffe F-4F Phantom II launched the missiles achieving direct hits on 25-cm (9.8-in) calibre Dornier SK6 target drones during head-on approaches after launching at off-boresight angles of +/- 50-degree and ranges of 5-km (2.7-nm). Test launches were conducted from a JAS 39 Gripen as well as other platforms. The first launches from the outboard wing stations on a Eurofighter Typhoon were conducted in April 2004 when Eurofighter DA7 conducted the missile launches at Decimomannu, Sardinia at an altitude of 5,000-ft at Mach 0.8 and 6-g and an altitude of 15,000-ft, Mach 0.7 and 6.6-g.

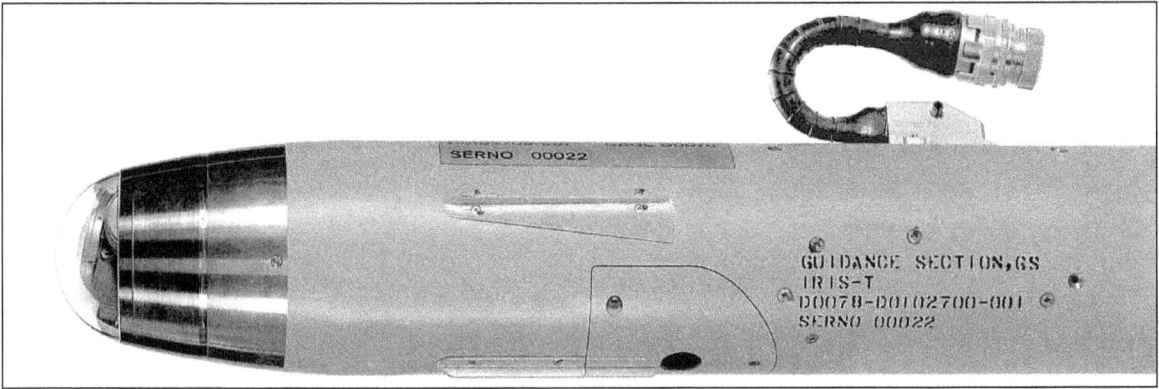

The heart of the IRIS-T capabilities stem from +/- 90-degree look angle high sensitivity seeker head which bestows upon the missile its high off-boresight capability.

Full clearance of IRIS-T via a digital interface was implemented in German, Italian and Spanish Tranche 2 Eurofighter Typhoons, which can carry up to four IRIS-T missiles across the entire flight envelope. Series production commenced in 2005 and the missile is now in service on Luftwaffe Eurofighter Typhoon and Tornado IDS, Italian and Spanish and Austrian Typhoons and the Swedish Saab JAS 39 Gripen. A total of five of the current Eurofighter Typhoon operators have selected the missile as its standard SRAAM, including the export customers, Austria and Saudi Arabia. The remaining Typhoon operator, the United Kingdom, equipped its Typhoons with the MBDA ASRAAM. South Africa and Thailand followed Sweden and selected IRIS-T to equip their respective JAS 39 Gripen fleets. IRIS-T is also being integrated on the F-16 in co-operation with Lockheed Martin, the weapon being ordered by Greece and the Spanish Air Force is also equipping its Boeing EF-18 Hornets with the weapon.

IRIS-T armed Austrian Typhoons escort an Airbus A380 on a photo shoot in 2011. Bundesheer

---

DIEHL BGT Defence IRIS-T

Propulsion: solid rocket motor
Diameter: 12.7-cm (5-in)
Fuses: active-laser proximity and contact
Guidance: passive IIR

The manufacturer states the main features of IRIS-T as the following:

-Roll-pitch imaging infrared seeker with a large look angle, most modern signal processing and high target tracking rate
-Seeker assisted radar proximity fuse
-Large warhead
-Motor optimised for dogfight
-Tail controlled, winged airframe with combined aerodynamic and thrust vector control
-Mass, length, diameter and centre of gravity similar to AIM-9L/M Sidewinder
-Fully compliant with existing analogue Sidewinder interfaces
-All-up round / certified round logistics concept

The highly agile IRIS-T is the only totally newly developed Short Range Air to Air Missile with an imaging IR seeker, dogfight optimised rocket motor, wings, and a combined aerodynamic and thrust vector control. The seeker assisted radar proximity fuse and the large warhead gives the missile a remarkable anti-missile capability. Predictive flight path tracking and lock-on-after-launch features enable the missile to engage targets in the rear hemisphere as well. The IRCCM and DIRCCM capabilities of IRIS-T are unmatched and the missile literally needs no maintenance. The IRIS-T is a state of the art fire-and-forget short range air to air missile with unrivalled close in combat and intercept performance. IRIS-T's outstanding capabilities and performance will totally change the nature of future air warfare.

Chapter Three

United States Air to Air Missiles

From the 1960's until the advent of the AIM-120 AMRAAM in the early 1990's, the AIM-7 Sparrow has been the standard US medium-range air-to-air missile, serving in a number of variants. A St Louis ANG Boeing F-15A Eagle interceptor launches an AIM-7M from a starboard wing inboard station stub pylon during a live fire missile training exercise. USAF

Raytheon AIM-7M Sparrow

The AIM-7 Sparrow has been in service with the USAF/USMC/USN for some five decades in a number of variants. Sperry and the USN originally developed the Sparrow as an air defence missile to arm fleet air defence fighters operating from USN aircraft carriers. Later variants of the missile were developed and produced by Raytheon, Hughes missile systems (now Raytheon) and General Dynamics. The current AIM-7M models are

extremely capable weapons compared to the early models fielded during the Vietnam War era, which were unreliable at best. The Sparrow is a semi-active radar homing weapon which is guided to the target by the launch aircraft's on-board radar before homing in on the target in the terminal phase. The AIM-7M has all-weather, all-aspect capability, able to track targets ranging from high-performance fighter aircraft to slow moving helicopters at low to high altitudes from any direction.

The Sparrow is made up of four main sections consisting of a guidance section at the front, warhead behind, control section and rocket motor at the rear. The cylindrical body has four mid-body wings and four smaller tail fins. Externally the Sparrow has remained more or less the same from the early models to today's AIM-7M; however, internally, changes to the missile systems have

introduced improvements to the missile guidance capability. The AIM-7M variants are highly manoeuvrable and can be used against close-range targets in a dogfight as well as against targets at BVR.

The Sparrow has been developed into a sophisticated air-to-air missile system compared with the relatively primitive and unreliable weapons employed on fighters in the 1960's. Early Sparrow missiles are shown with the McDonnell Douglas F4H-1 Phantom II in the early 1960's. Boeing

transformed Sparrow from a relatively unreliable weapon into a potent air defence system.

The AIM-7F entered service with the USAF in 1976 as the primary air-to-air armament of the McDonnell Douglas (now Boeing) F-15A/B Eagle air defence fighter. The improved AIM-7M entered USAF service in 1982. This variant featured improved reliability and increased performance over the earlier Sparrow variants, primarily at low-altitudes and in a dense electronic countermeasures environment. In addition, the AIM-7M was armed with a more lethal warhead. In 1987, the H-Build software release was introduced and this remained the standard release in the early 2000's. This

Top: The AIM-7M Sparrow is retained by USANG (United States Air National Guard) F-15 Eagle squadrons tasked with the defence of the Continental United States from air attack. Post September 11, 2001, the interception of civilian airliners has increased. A Sparrow is launched from a Florida ANG F-15A. Above: An AIM-7M is launched from a USAF F-15E Eagle. Both USAF

The AIM-7M was used operationally by USAF F-15 Eagles during the January to March 1991 Persian Gulf war against Iraq where it was credited with shooting down twenty two Iraqi Air Force fixed wing aircraft and three Iraqi helicopters.

In USAF service the AIM-7M can be carried by the F-15A/B/C/D/E Eagle and the Lockheed Martin F-16C/D. The USN/USMC operate the missile from Boeing F/A-18A/B/C/D Hornet strike fighters, while the Sparrow also formed the primary air-to-air armament of the Northrop Grumman F-14A/B/D Tomcat strike fighter in USN service until that aircraft was retired in 2007

The USN also employs a surface launched variant, designated RIM-7M NATO Sea Sparrow, with the primary function of surface to air anti-ballistic missile defence. Improvements to the surface launched system led to the ESSM (Evolved Sea Sparrow Missile), which was an international upgrade program for the RIM-7. The first ESSM systems were delivered to the USN in 2002. The ESSM provides the primary air defence capability for ten NATO navies.

Raytheon AIM-7M Sparrow

Propulsion: Hercules MK-58 solid-propellant rocket motor
Length: 3.64-m (12-ft)
Diameter: 0.20-m (8-inches)
Wingspan: 1-m (3-ft 4-in)
Launch weight: approximately 225-kg (500-lb)
Warhead: annular blast fragmentation
Guidance system: Raytheon semi-active on either continuous wave or pulsed-Doppler radar energy
Date deployed: 1982

Top left: From the early 1980's when it entered operational service with the USN/USMC, the McDonnell Douglas (now Boeing) F/A-18 Hornet enjoyed a BVR air to air capability among the best in the world courtesy of its advanced Hughes (now Raytheon) AN/APG-65 multi-mode radar system and the Hughes (now Raytheon) AIM-7M Sparrow missile. Boeing **Centre top and centre bottom left: By comparison, the General Dynamics (now Lockheed Martin F-16 Fighting Falcon lacked a BVR air-to-air capability before the introduction of AMRAAM. The exception to this was the F-16A/B ADF (Air Defence Fighter) variant, which was purchased for the USANG for the air defence of the Continental United States. These aircraft have been withdrawn and replaced by newer F-16C/D's armed with AMRAAM and F-15 Eagles armed with AMRAAM and Sparrow.** Both LMTAS **Left: AIM-7 Sparrow missiles were also supplied to foreign customers, mainly F-15 Eagle operators. A RSAF F-15 Eagle launches an AIM-7F Sparrow.** Raytheon

An AIM-120 AMRAAM is launched from an USAF Boeing F-15C Eagle over a missile range in the US. The F-15 Eagle was the first platform to be armed with AMRAAM in 1991. Raytheon

Raytheon AIM-120 AMRAAM

Designed as a direct replacement for the AIM-7 Sparrow semi-active radar homing missile, the AIM-120 AMRAAM, which is smaller, lighter and faster than its predecessor, has revolutionised NATO air-to-air capability. Unlike the vintage Sparrow, which required the launch aircraft to continually keep the target illuminated by its radar during the missile flight time, the AMRAAM has its own active-homing radar on board, together with an INS (Inertial Navigation System) and a datalink. Once launched, AMRAAM will fly towards the target controlled by the pre-programmed inertial guidance system, requiring no further assistance from the launch aircraft. For longer-range engagements the operator can provide mid-course guidance, transmitting target location data to the missiles guidance system, following which the missile goes into the autonomous mode, using inertial guidance only. When it nears the target area it switches to the terminal homing mode using the missiles on-board active monopulse radar seeker. AMRAAM can be used to engage multiple targets simultaneously whereas with the Sparrow or British Skyflash only one target could be engaged at a time.

As well as improvements in capability over the SARH AIM-7M Sparrow, AMRAAM was also more versatile in its application to airborne platforms and in the weapon stations which could accommodate the missile. An AMRAAM is carried on the inboard wing station stub-pylon of an F-15E Eagle, previously associated with the AIM-9 Sidewinder SRAAM. Author

Top: With the introduction of a mid-life upgrade for their respective F-16 fleets, European F-16A/B operators introduced AMRAAM to service giving the aircraft a BVR capability for the first time. This Dutch F-16AM is landing with AMRAAM's on the wingtip stations. **Above:** A UAE F-16F Block 60 lifts of with AMRAAM missiles on the wingtip launch stations. Both LMTAS

Previous page top: The Lockheed Martin F-16 CFT (Conformal Fuel Tank) development aircraft is seen during a test flight in 2000 carrying AIM-120 AMRAAM's on the wingtip missile launch rails. LMTAS Previous page bottom: An USAF F-16C Fighting Falcon launches an AIM-120 AMRAAM from a port side under wing station. USAF

Top: A USAF Lockheed Martin F-22 launches an AMRAAM from the aircrafts port side internal weapons bay. Above: Although the F-22 Raptor can carry the AIM-120C internally AMRAAM's and other stores can also be carried externally as seen on this Raptor with AMRAAM's on the outboard under wing stations. External carriage of stores would obviously impact on the low observable characteristics of the aircraft. LMTAS

Two USAF Boeing F-15E Eagle strike fighters conduct multiple launches of AIM-120 AMRAAM missiles during a live fire launch over a US range-facility. Despite being fielded primarily as a strike aircraft the F-15E armed with AMRAAM is one of the most capable air combat fighters fielded by any air force in the 1990's. USAF

AMRAAM was designed in the 1970's and the conceptual phase of the program was completed in February 1979 when two of five competing contractors (Hughes and Raytheon) were selected by the USAF to continue to the validation phase. The 33-month validation phase included both contractors producing hardware to support their respective concepts. In December 1981 the validation phase of the program was concluded when both contractors apparently demonstrated that their respective flight-test missiles could meet the USAF and USN requirements. The USAF as the full-scale developer selected Hughes Missile Systems group and Raytheon was selected as a follower producer.

In 1987 Hughes and Raytheon received production contracts for AMRAAM missiles. Hundreds of missiles were test-launched at Elgin AFB, Florida, White Sands Missile Range, New Mexico and Point Mugu, California.

A British Aerospace (now BAE Systems) Sea Harrier FA.2 launches an AIM-120B AMRAAM during development integration. Raytheon

An USAF F-16 launches an AMRAAM during a test firing. Integration of AMRAAM turned the F-16 into a truly multi-role strike fighter with an impressive BVR air-to-air capability to augmenting its already impressive short-range air combat capability. USAF

AMRAAM entered service in time to be employed operationally during the 1991 Gulf War; however, no missiles were launched before the cease-fire. AMRAAM has been used in several engagements by US and NATO forces against Iraqi and Serbian aircraft since 1992 with mixed results. The first combat use of AMRAAM occurred on 27 December 1992; a USAF F-16D Block 42 fighter launched an AMRAAM from BVR at an Iraqi MiG-25 'Foxbat' E in the controversial no fly zone over Southern Iraq, initially set-up by France, the UK and US, but not sanctioned by the United Nations. The missile destroyed the Iraqi fighter.

AMRAAM has a similar configuration to Sparrow with four fixed central wings and four rear control fins. The missile is smaller and lighter than its predecessor with a length of 3.65-m, a diameter of 17.8-cm, a span of 63-cm, and weighs in at 157-kg (345.4-lb), including the 23-kg (50.6-lb) pre-

fragmented High Explosive warhead with either a proximity or contact fuse. Powered by a solid propellant rocket motor the missile has a speed of Mach 3 and a maximum range of 65-km (40.4 miles), being faster and having a longer engagement range than its predecessor and is also more manoeuvrable and is more resistant to ECM. As long as the launch aircraft has a track while scan radar up to six missiles can be fired in rapid succession against separate targets.

This video still shows the first launch of an AMRAAM from Eurofighter DA7 in 1997. Eurofighter GmbH

Top: The Saab JAS 39 Gripen was developed with the AIM-120 AMRAAM as the standard BVR air-to-air weapon for the Flygvapnet. The fifth prototype Gripen launches an AMRAAM during weapons integration testing. Saab **Above: F-35A, AF-1, conducts the types first in-flight missile launch, an AIM-120C5 AAVI (AMRAAM Air Vehicle Instrumented), round in June 2013.** Lockheed Martin

Prior to the introduction of Meteor the AIM-120 is the primary BVR air-to-air weapon of all four Eurofighter Typhoon partner nations and will remain in service with Typhoons until the Meteor is well established in service. Eurofighter GmbH

The Pre-Planned Product Improvement (P3I) program was designed to ensure that AMRAAM capability was sustained throughout its service life. The first two phases included the clipped wings and fins for the AIM-120C to allow it to be carried internally in the Lockheed Martin Boeing F-22 Raptor and the Lockheed Martin F-35 Lightning II and an improved warhead and fuse enhanced ECCM (Electronic Counter Counter Measures) capability and a reduced length, SCAS (Shortened Control and Actuation Section). In 1999 a 5-in longer rocket motor that utilised the space made available with the SCAS was put into production. These improvements would be implemented into new production missiles, but the designation would remain unchanged, with Block numbers identifying modification standard. The P3I Phase III contract was awarded in November 1998 and this would introduce further improvements to the basic AMRAAM, which has a reported shelf life of 25 years. The missile has a MTBF (Mean Time Between Failure) of 1,334 hours on USAF Boeing

F-15 and Lockheed Martin F-16's and 888 hours on Boeing F/A-18 Hornet strike fighters.

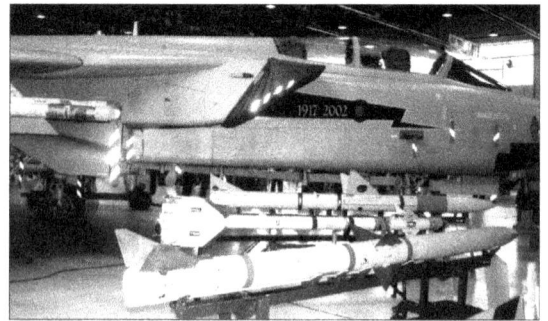

As well as the Typhoon the UK operated AMRAAM with the Sea Harrier top right and the Tornado F.3 CSP fighter above (both now retired). Author

Previous page top: An AMRAAM is launched from an USAF F-16C. USAF Previous page bottom: An AMRAAM is shown on the starboard forward fuselage missile recess of a German Air Force Typhoon. Eurofighter GmbH Above: An AMRAAM rocket engine ignites after it is launched from the fuselage internal weapons bay of a Lockheed Martin F-22 Raptor air dominance fighter. USAF

When introduced to service AMRAAM was rightfully regarded as a quantum leap in capability over the AIM-7 Sparrow that it was replacing. However, while the missile has been successfully used during a number of operations and its capabilities demonstrated in numerous test firings, there has long been concern over an apparent drop-off in speed during long-range flights following rocket motor burn out. As the speed falls off the missile becomes less capable of intercepting a manoeuvring target. After launch, during rocket motor burn, AMRAAM accelerates to Mach 3, but following rocket motor burn out speed gradually decreases to between Mach 1 to Mach 1.5. The deficiencies were highlighted in January 1999

when a number of missiles were fired at Iraqi MiG's during two separate engagements, all of which failed to achieve a kill. It is thought that as speed fell off the Iraqi MiG's simply out ran the missiles.

An AIM-120 AMRAAM rocket motor ignites as the weapon launches from the port side under wing station stub-pylon of a USAF F-15 during a missile firing over a test range in the US. USAF

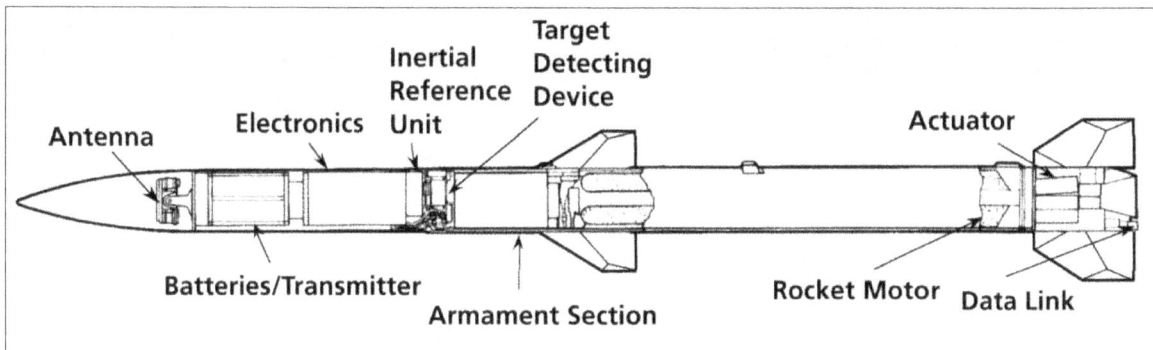

A cutaway drawing of a Raytheon AIM-120 AMRAAM shows the position of all of the weapons major components. Raytheon

To address the problem AMRAAM demonstrations conducted with improved rocket motors. However, this problem, combined with the emerging threat of longer range advanced air to air missiles from Russia have seen most Western missile manufactures looking at ramjets for their respective BVR missiles. During the early 1990s, there was USAF interest in a ramjet-powered AIM-120 that would maintain the Mach 3 speed all the way to the target, providing up to four times the 'no escape zone' of comparable rocket powered missiles like the standard AMRAAM.

Top left: An USAF F-16 launches an AMRAAM. Above: A USAF F-15 banks to starboard showing the carriage of AMRAAM missiles on the fuselage shoulder stations. Both USAF

ARC (Atlantic Research Corporation) and Alliant Techsystems began working on a VFDR (variable-flow-ducted-rocket) in 1985 and in 1997 ARC revealed that they had ground tested a VFDR which demonstrated a successful transition from rocket powered flight to the solid fuel ramjet sustainer. The Pentagon, however, announced that it had no plans to develop an extended range AIM-120.

AMRAAM can operate in three modes, depending on target range and conditions of engagement. Once launched flies towards the target under control of the pre-programmed inertial guidance system, needing no further assistance from the launch aircraft. For long-range engagements the aircraft can update the missile flight path mid-course, transmitting target location data to the missile guidance section, following which it goes into autonomous mode, using inertial guidance only. Once near the target it switches to terminal homing mode using the on board active monopulse radar seeker.

In the late 1980's it was expected that the Grumman F-14 would be replaced by a naval variant of the USAF ATF (Advanced Tactical Fighter) contest won by the F-22. While the USAF was focusing on the AIM-120 as the principal air-to-air missile for its ATF, the USN wanted a longer-range replacement for its Hughes AIM-54A/C Phoenix. In 1988 the USN awarded development contracts for an Advanced Air to Air Missile (AAAM) to 2 design teams consisting of Hughes / Raytheon and General Dynamics / Westinghouse. However, the AAAM, which was to be lighter than the Phoenix, but with a longer range, was cancelled as the navalised ATF fell by the wayside. With cancellation of the AAAM the USN BECAME reliant on the AIM-120, which while being an excellent air-to-air missile is considered by the USN to have an inadequate engagement range for many of the BVR scenarios envisaged for the 21st century.

A USMC Boeing F/A-18D Hornet strike fighter is armed with AMRAAM and Sidewinder air-to-air missiles during an operational mission during the 1999 NATO bombing campaign over Serbia and Kosovo. USMC

AMRAAM has been integrated with a number of US and foreign fighter aircraft including the Boeing F/A-18A/B/C/D Hornet, Boeing F/A-18E/F Super Hornet, Boeing F-15A/B/C/D/E Eagle, Lockheed Martin F-16AM/BM/C/D/E/F Fighting Falcon, Lockheed Martin F-22 Raptor, Boeing/BAE Systems AV-8B Harrier II+, Boeing F-4 Phantom II, Eurofighter Typhoon, RAF Panavia Tornado F.3 (now retired), BAE Systems Sea Harrier F/A.2 (now retired) and Saab JAS 39 Gripen.

AMRAAM has been used in a number of operational theatres since it entered service in 1991. Two Iraqi Air Force aircraft were shot down over the controversial southern No Fly Zone over Southern Iraq and one AMRAAM was used to shoot down a Bosnian Serb aircraft over Bosnia in 1995. During operation Allied Force – the NATO bombing campaign against Serbia and Montenegro/Kosovo from March to June 1999, US and NATO aircraft used AMRAAM to shoot down a number of Serbian fighters.

Up to 2012, AMRAAM's customer base continued to grow with orders coming from several nations, which included new customers such as Pakistan and Taiwan and continuing customers such as Singapore while the US services continue to order new batches. 2012 also saw further problems emerge with the weapon as deliveries to Finland were withheld apparently due to problems with the weapons when operated in very cold temperatures such as those encountered in Northern Finland. That said; several other customers such as Sweden and the USAF have operated the weapon in cold environments. As of 2012 AMRAAM is operated by no less than 39 nations on a variety of combat aircraft types.

Raytheon AIM-120 AMRAAM

Propulsion: solid propellant rocket motor
Length: 366-cm (12-ft)
Diameter: 17.78-cm (7-inches)
Wingspan: AIM-120A/B (52.58-cm/21-inches), AIM-120C (12.5-inches)
Launch weight: 150-175-kg (335-lb)
Speed: Mach 3
Range: 17.38-nm (20+ miles) according to USAF data, although other data claims a range of around 40-50-km
Warhead: high explosive blast fragmentation
Guidance system: active-radar terminal inertial mid-course

The AIM-9 Sidewinder short-range infrared guided air-to-air missile, in its many variants, has been the most widely used air-to-air missile in history. The AIM-9M carried by this USN Boeing F/A-18C Hornet is an extremely capable air-to-air weapon with all-aspect engagement capability. USN

Raytheon AIM-9 Sidewinder

The AIM-9 Sidewinder, which is an infrared guided missile designed for within visual range air to air combat, has undergone continued development since the first primitive models entered service in the late 1950's; originally developed by the USN to arm fleet-air-defence flghters like the Vought F-8 Crusader. The USAF subsequently adopted the weapon and it became the standard IR guided air-to-air weapon in US service.

The Sidewinder has a cylindrical body made-up of a rocket motor section, a high explosive warhead section and an active optical tracking system section. At the rear of the missile is a roll-stabilising rear wing/rolleron assembly. Manoeuvrability is increased by the addition of detachable double-delta control surfaces located behind the nose section.

The infrared guidance gives the missile its fire-and-forget capability, allowing the launch aircraft to leave the area or take evasive action after launch. The IR guidance seeker head homes in on the heat signature generated by the target aircraft particularly the engine exhaust.

Top right: The AIM-9B and other early variants entered service from the late 1950's, serving on a number of platforms including the USN F9F. USN
Above: The Sidewinder was widely exported serving on a number of US built and foreign fighter aircraft including the Swedish Saab JA 35 Draken. Saab

Top and above: In the 1970's and 80's a number of new build Sidewinder variants and upgrades of older weapons were developed including the AIM-9J/P models seen on an F-16C (top) and a Greek F-16D Block 50+ (bottom). Both LMTAS

This USN Northrop Grumman F-14 Tomcat from VFA-32 is armed with an AIM-9M on the port wing station outer stub-pylon. USN

A prototype of the AIM-9A was successfully launched in September 1953. This paved the way for the initial production variant, the AIM-9B, which was delivered to the USAF from 1956. This variant, however, was extremely limited in capability; being effective only at very close-range, it could not be used at night time, it lacked head-on engagement capability and could not cngagc targets flying close to the ground. Some of the later early models increased the missile effectiveness. The AIM-9J variant was an upgrade of the AIM-9B and AIM-9E stocks. This variant introduced manoeuvrıng capabılıty for close-range dogfighting and also increased the missile's speed and range. The USAF began receiving AIM-9J's to arm McDonnell Douglas F-15A/B Eagle fighters in 1977.

The AIM-9L variant introduced a manoeuvring tracking capability and a more powerful solid propellant rocket motor. Capability and resistance to countermeasures was further increased by the introduction of an improved active-optical fuse.

The increased sensitivity conical-scan-seeker improved tracking stability. The most important improvement in capability for the AIM-9L over earlier generation Sidewinders was the all-aspect capability (able-to attack the target from any direction). Production of the AIM-9L began in 1976 and deliveries to the USAF commenced that same year.

An AIM-9M is launched from the starboard wingtip launch rail of a USAF Lockheed Martin F-16. LMTAS

Boeing F/A-18D Hornet strike fighters from the USMC VMFA(AW)-533 armed for an air to surface mission with AIM-9M Sidewinders carried on the wingtip stations for self-defence during the 1999 NATO bombing campaign over Serbia and Kosovo. USMC

The AIM-9P, deliveries of which commenced in 1978, was developed as an upgrade of the AIM-9J featuring greater engagement boundaries over its predecessor, including the ability to be launched further from the target. The new variant was also more manoeuvrable and included solid-state electronics increasing reliability and maintainability. The AIM-9P-1 replaced the infrared-influence-fuse with an active-optical target detector, while the AIM-9P-2 introduced a reduced smoke rocket motor. A later variant of the 'P' combined the active optical target detector with the reduced smoke engine. This later variant also had mechanical strengthening to the warhead. The new warhead used explosive, which was less sensitive to heat.

From 1960, the AIM-9B was produced by BGT in Europe under a licence production agreement with the US. Around 10,000 AIM-9B missiles were produced under BGT leadership and delivered to eight European NATO air forces. BGT went on to develop the FGW Mod 2 variant from the AIM-9B. This variant was developed to be better capable of

operating in the adverse weather conditions of a North European environment and had a better performance at low-altitudes. By 1973, around 7,000 missiles of this variant were produced for service with a number of European air forces.

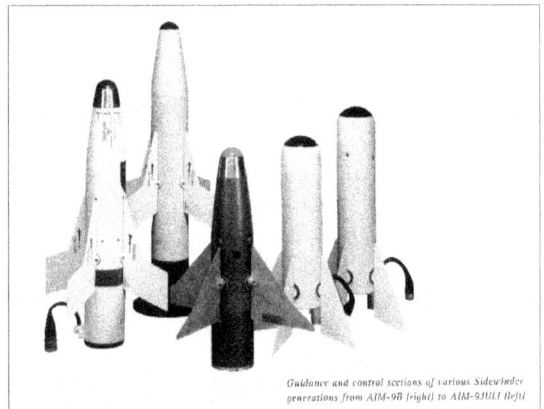

Guidance and control sections of various Sidewinder generations from AIM-9B (right) to AIM-9JULI (left)

The AIM-9 has been licence produced in Europe since the 1970's. A number of new build variants and upgrades of older missile stocks have been undertaken with BGT in Germany as prime contractor. The guidance and control sections for a number of models over the generations are shown with the AIM-9B at far right, AIM-9L/I (centre) and AIM-9JULI at far left. BGT

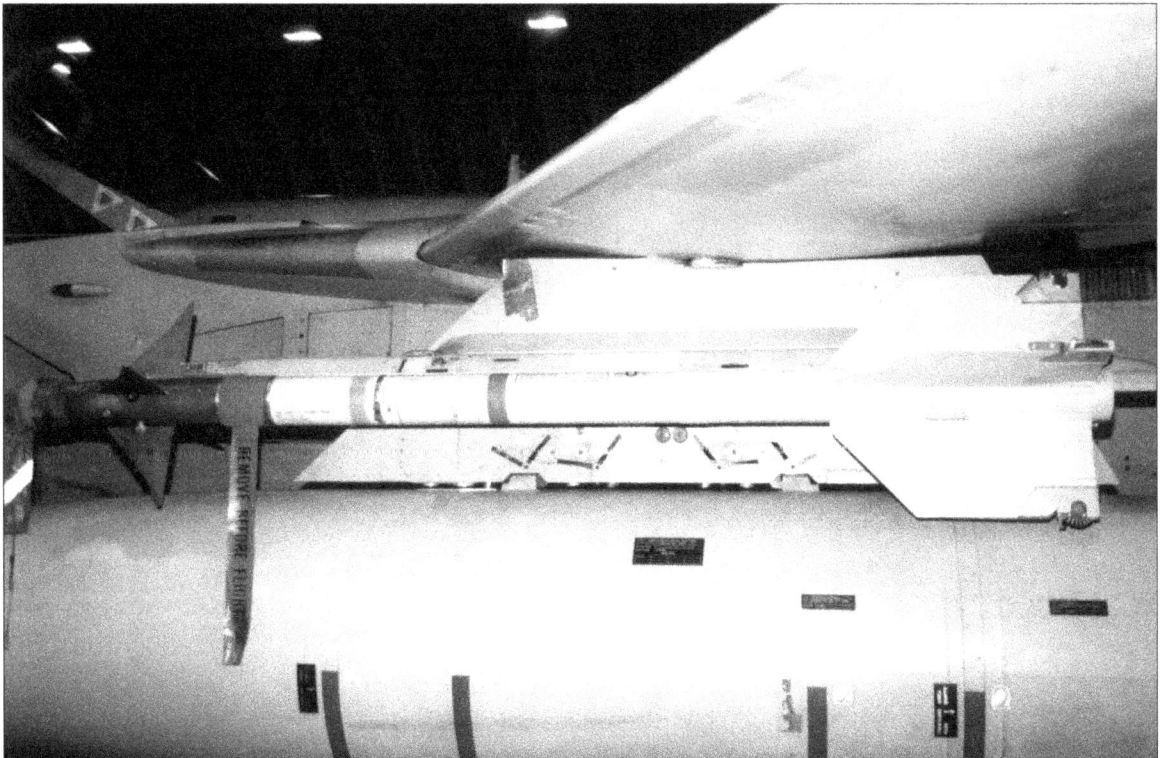

When it entered service in the mid-1980's the RAF Panavia Tornado F.3 fighters were armed with AIM-9L Sidewinders. Top: An F.3 lets loose an AIM-9L during a missile practice camp in the late 1980's. BAE Systems **Above: A live AIM-9L is carried on the port wing station outer stub-pylon of a RAF Tornado F.3 on QRA (Quick Reaction Alert) duty at RAF Leuchars in 1993.** Author

The Eurofighter Typhoon was cleared with the AIM-9L variant of the Sidewinder for initial service with Germany, Italy and Spain. RAF Typhoons have also been cleared with the AIM-9L, however, that services Typhoons were armed with **MBDA ASRAAM as the primary IR guided air-to-air missile. Top: Germany's Eurofighter DA5 lands with a pair of AIM-9L's on** the outboard wing stations. Above: The third production Typhoon, IPA3, is seen in flight with **AIM-120 AMRAAM's on the fuselage shoulder stations and a pair of AIM-9L's on the outboard wing stations.** Both Eurofighter GmbH

The AIM-9M is currently the only variant of the original Sidewinder family still in service with USAF, USN and USMC fighter squadrons. Top: A pair of F-16C fighters from the 21st FS, Luke AFB, fly close formation with AIM-9M's on the wingtip launch rails. USAF Above: A USN F/A-18C from VFA-94 is positioned on the carrier deck with AIM-9M's on the wingtip stations. USN

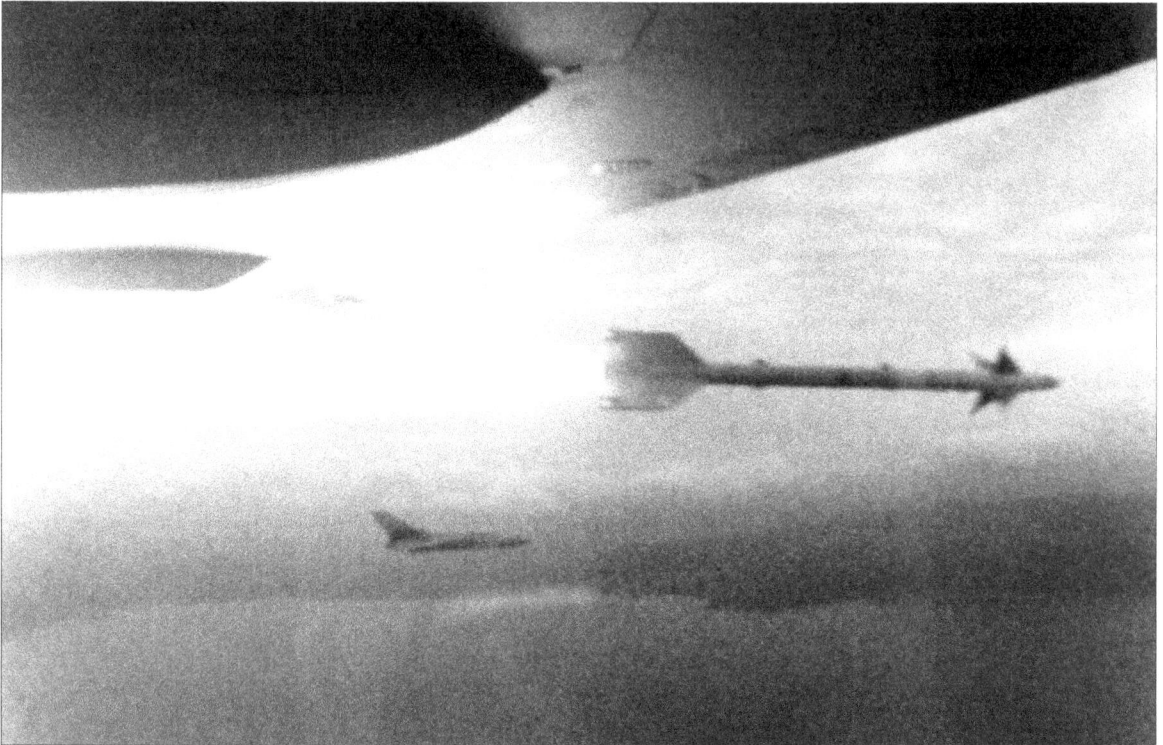

Above and right: The Italian Eurofighter development aircraft DA7 launches an AIM-9L Sidewinder IR guided air to air missile for the first time on 15 December 1997, beginning that aircrafts clearance for deployment of the missile with in service Typhoons from 2003. Eurofighter GmbH

From the mid-1970's, BGT became the prime contractor for the European AIM-9L production program, which was a four-nation program consisting of the then West Germany (now Germany) as lead nation, Italy, Norway and the United Kingdom. Around 15,000 AIM-9L missiles were produced for nine European nations by the end of the 1980's. In the early 1980's, West Germany, Italy and Norway went on to order an improved AIM-9L known as the AIM-9L/I, whose seeker-head was capable of suppressing the range of flare countermeasures then known in the West. The main improvement consisted of a newly developed guidance electronics module in hybrid technology. Most of the AIM-9L stocks in Europe were eventually retrofitted.

The AIM-9L/I-1 further increased resistance to IR countermeasures. This variant is claimed to be resistant to all-known countermeasures to date and is more resistant to the more recent IRCM. It is thought that many European air forces will upgrade at least part of their existing AIM-9L/L/I stocks to this standard, although the introduction of more

modern missiles such as IRIS-T ultimately reduced the potential market.

A USAF F-16D launches an AIM-9M from the port side wingtip station. USAF

Top: The Lockheed Martin F-22 Raptor has been developed as the USAF premier air combat fighter for service in the first few decades of the 21st Century. The Raptor has been cleared to use the AIM-9M variant of the Sidewinder, which is housed in two weapons bays – one on each of the fuselage sides. This was the first supersonic speed missile launch from an F-22, conducted on 25 July 2002. USAF Although the AIM-9X is in operational service the AIM-9M remain part of the armament options for tactical fighters such as USAF Lockheed Martin F-16 Fighting Falcon (above). Raytheon

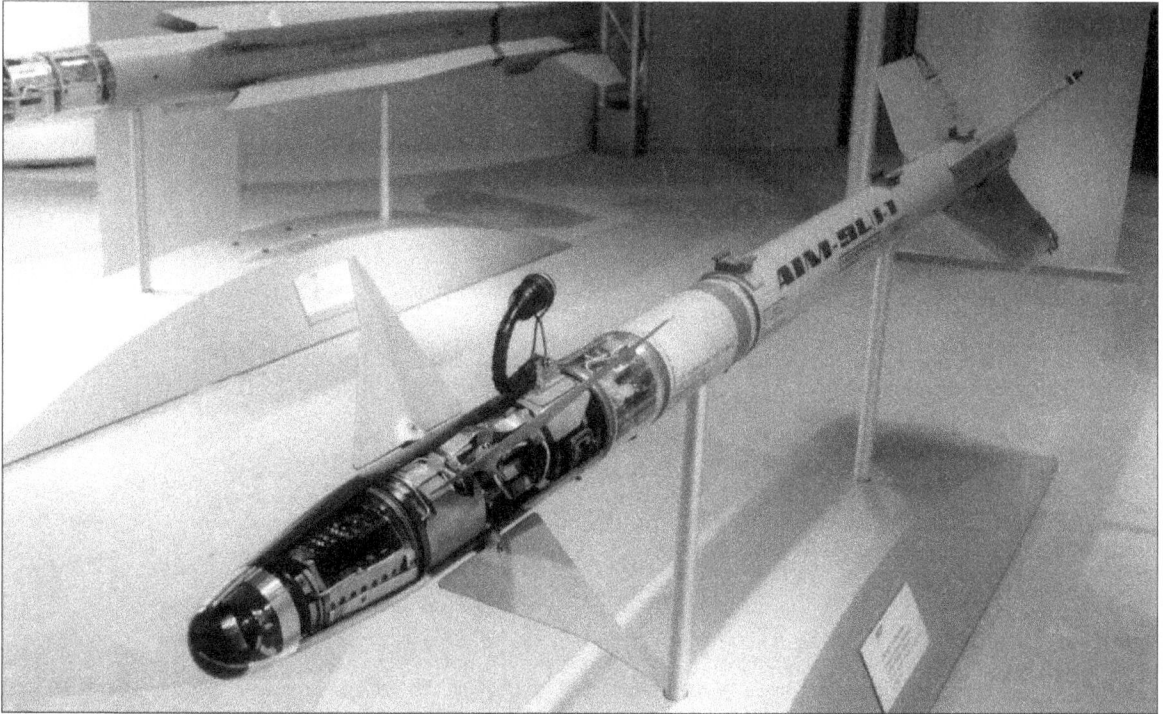

The AIM-9L/I-1 was marketed by BGT as an AIM-9L upgrade. Author

BGT also developed a number of upgrades for older AIM-9N/P stocks to enable them to engage targets from the forward hemisphere and increase resistance to countermeasures. The BGT developed AIM-9JULI, which replaced the AIM-9N/P seeker head with the increased capability seeker and guidance electronics from the AIM-9L/I or -9L/I-1. The AIM-9JULI is a cost effective upgrade for air forces operating large stocks of older AIM-9N/P variants of the Sidewinder. This upgrade has already been delivered in large quantities.

The most recent US variant of the original Sidewinder is the AIM-9M, which retained the all-aspect capability of the AIM-9L, but featured increased all-round performance. Defence against counter measures was increased and the missile featured enhanced background discrimination capability and a reduced smoke rocket motor. The improvements increased the missile's ability to locate and lock-on to a target and deliveries of the AIM-9M commenced in 1983. A later variant, the AIM-9M-9 introduced expanded infrared countermeasures detection circuitry.

The AIM-9 has been integrated with a large number of aircraft types in a large number of air arms around the world. Aircraft types integrated with the missile include the Boeing (formerly McDonnell Douglas) F-4 Phantom II, F-15A/B/C/D/E/S/I/K Eagle, F/A-18A/B/C/D/E/F/ Hornet and Super Hornet, A-4 Skyhawk, the Boeing/BAE Systems AV-8B Harrier II, Harrier GR.5/7/9/10/12, the Lockheed Martin F-16A/B/C/D/E/F Fighting Falcon and Desert Falcon, F-22 Raptor, Northrop Grumman F-5, F-14 Tomcat, A-6, Vought A-7, Fairchild A-10, Panavia Tornado ADV/ECR/IDS, Eurofighter Typhoon, Saab JA 35 Draken, JA 37 Viggen, JAS 39 Gripen, SEPECAT Jaguar, BAE Systems Hawk and BAE Systems Sea Harrier. The Sidewinder has also been adapted for use on attack helicopters like the Bell AH-1W Super Cobra.

Raytheon AIM-9M Sidewinder

Propulsion: Hercules and Bermite Mk 36 Mod 11 rocket motor
Length: 2.9-m (113-in)
Diameter: 12.7-cm (5-in)
Fin span: 63.5-cm (25-in)
Canard: 56.6-cm (22.3-in)
Launch weight: 86-kg (190-lb)
Warhead: 11.3-kg (25-lb) annular blast fragmentation
Fusing: proximity and contact
Guidance: solid-state passive IR homing system

An USAF F-16C Block 40 fighter armed with air to surface stores and AIM-120 AMRAAM air to air missiles on the wingtip stations and AIM-9X Sidewinder IR guided air to air missiles on the outboard under wing stations. USAF

Raytheon AIM-9X Evolved Sidewinder

The AIM-9X program developed a short-range heat seeking weapon to be employed in both offensive and defensive counter-air operations. Offensively, the weapon will ensure that US and combined air forces have the ability project the necessary power to insure dominant manoeuvre. In the defensive counter-air role, the missile system will provide a key capability for force protection.

The AIM-9X development fielded a high off-boresight capable short range heat seeking missile to be employed on US Air Force and Navy/Marine Corps fighters. The AIM-9X will complement longer-range radar guided missiles such as the Advanced Medium Range Air-to-Air Missile (AMRAAM).

On 13 December 1996, the US Naval Air Systems Command selected the Hughes (now Raytheon) Evolved Sidewinder, as its preferred next generation AIM-9X close range air-to-air missile. This choice was a surprise to many observers as it was less advanced than the other main contenders put forward by Raytheon and the Hughes/BAe Dynamics ASRAAM P3I. Hughes claimed that its Evolved Sidewinder combined

conformal performance with inherent growth capability in a low cost, low risk program.

The Evolved Sidewinder, also known as the Box Office 2 plus design, was developed under an USN demonstration/validation contract awarded in 1994. This weapon retained the AIM-9M's rocket motor, warhead and fuse. Using thrust-vector control increased manoeuvrability and the missile has a new IR seeker head.

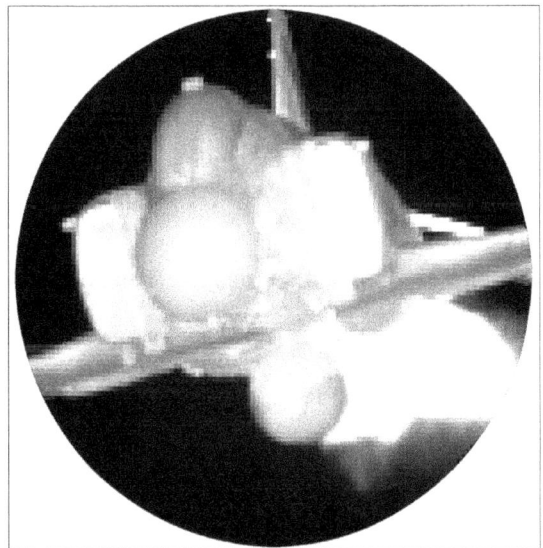

View of a target as seen from the AIM-9X IR seeker-head. Raytheon

An AIM-9X is launched against a QF-4N target drone by an F/A-18 Hornet from the NAWC-WD (Naval Air Warfare Centre-Weapons Division) over the China Lake, California range in 1999. NAWC-WD

The new missile was required to re-establish some degree of parity of US aircraft in short range air combat, vis-à-vis improved foreign export aircraft and missiles. Specific deficiencies existed in the current AIM-9M in high off-boresight angle capability, infrared counter-countermeasures robustness, kinetic performance, and missile manoeuvrability. The Su-27/MiG-29 with their R-27 and R-73 missiles was seen as the major threat to US aircraft. Additionally, there were a number of other missiles on the world market that outperformed the US AIM-9M in the critical operational employment areas. The AIM-9X expands the capabilities of the AIM-9M by incorporating a new seeker imaging infrared focal plane array, a high performance airframe, and a new signal processor for the seeker/sensor. The acquisition strategy set out to retain the warhead,

fuse, and rocket motor of the AIM-9M in order to capitalise on the large existing inventory of AIM-9 weapons.

A number of evaluations of the competing missiles systems put forward to meet the AIM-9X requirement took place prior to selection of a winner. The three main contenders received the most rigorous evaluation; these being the two US AIM-9X DEM-VAL participants and the British ASRAAM. The purpose of the operational test agency's efforts was to assess the potential operational effectiveness and suitability of the systems under consideration to support a 2 December 1996 Defence Acquisition Board (DAB) and down-selection to one final contract award. The assessments were based upon laboratory, simulation, and captive test unit data collected during the developmental effort and dedicated foreign comparative testing.

The EOA (Early Operational Assessment) of the Hughes and Raytheon DEMVAL (Demonstration Validation) results was executed by COMOPTEVFOR. The EOA focused on four risk

Although retaining many components from the AIM-9M, the AIM-9X emerged as a much aerodynamically sleeker looking missile with smaller control surfaces. Raytheon

areas of missile development: focal plane array performance, detector cooling, gimbal and platform stabilisation, and signal processing. Data gathering activity in two flight phases (15 ground-to-air and 37 air-to-air sorties) and a co-ordinated modelling and simulation effort supported the assessment.

An AIM-9X is carried on the starboard wingtip launch rail of an USN Boeing F/A-18F Super Hornet strike fighter. Author

The flight phases assessed the missile performance in various background, countermeasure and target aircraft manoeuvring conditions. The result of the assessment was that both the Hughes and Raytheon missiles showed

potential for meeting both the mission effectiveness and suitability requirements of the AIM-9X operational requirement document. Specifically, all critical operational issues were rated green (potentially effective/suitable) except counter-countermeasures capability, lethality, built in test functionality, and reprogram ability. Counter-countermeasures capability of both missiles was initially below the operationally required threshold values; however, the Hughes missile showed a rapid improvement through the course of the evaluation.

The missiles demonstrated acceptable performance levels in the air-to-air phase. The other assessment areas not resolved as green had insufficient data for conclusive evaluation. However, again, the risk of either DEMVAL missile not meeting the threshold requirement was rated as low. The results of the operational assessment were integral to the Service source selection decision to award the engineering, manufacturing, development contract to Hughes Missile Systems Corporation. The early operational assessment of the British ASRAAM FCT (Foreign Comparative Test) focused on the risk areas of the ASRAAM: focal plane array effectiveness, seeker signal processing, warhead effectiveness, rocket motor testing, and kinematics/guidance ability to support the lethality requirements of the AIM-9X. The initial FCT plan called for nine ground-to-air test flights, 40 captive carry air-to-air flights, four programmed missile launches, warhead testing, and motor case testing.

The AIM-9X evolution from the AIM-9M featured a
new high-off-boresight IR seeker-head, new forward
fixed wings (top) and new much smaller tail control
fins and jet vein control at the rear (above). Author

An AIM-9X is launched from an Edwards Air Force Base; California based USAF Boeing F-15D Eagle. Raytheon

After several modifications to the scope of the FCT the EOA assessed: four ground-to-air sorties, 19 air-to-air captive carry sorties, four programmed missile launches, eight static warhead tests and four rocket motor case tests. Due to immaturity of the British furnished modelling and simulation tools and missile design, AFOTEC was not able to individually report on each critical operational issue. The assessment, instead, aggregated the issues in three major assessment areas: seeker/tracker performance, airframe kinematics capability, and mission capability/system maturity. The resulting assessment was that the ASRAAM was deemed to not meet the AIM-9X operational requirements in high off-boresight angle performance, infrared counter-countermeasures robustness, lethality, and interoperability.

Following its selection as the AIM-9X choice Hughes (now Raytheon) received an EMD contract valued at $169 million and funding was provided for the first 1,000 missiles with up to 10,000 units being envisaged. Initial operational capability was achieved in 2003.

The AIM-9X successfully completed its first guided launch by intercepting a QF-4N drone on 30 June 1999 at the NAWC-WD China Lake ranges in California. The launch, from a NAWC-WD NWTS (Naval Weapons Test Squadron) F/A-18D Hornet, which was the first in a series of guided launches off F/A-18 and F-15 aircraft, marked a significant milestone for the AIM-9X program as it was the first guided launch evaluating the ability of the missile's high off-boresight infrared seeker to track a full-size target through launch, fly-out and intercept. The specific launch objectives were to demonstrate airframe controllability through separation, demonstrate tracking post launch transient and guidance initialisation and guide to an aim point on the target aircraft hard body up to the terminal transition point.

On 3 March 2000 the AIM-9X successfully completed its fourth guided launch, intercepting a QF-4 at China Lake. This launch, from an F/A-18C was the first launch conducted in a dogfight scenario in the presence of defensive infrared countermeasure flares as well as being the first launch in conjunction with the JHMCS (Joint Helmet Mounted Cueing System). A few weeks later on 21 April 2000, an AIM-9X launched from an F/A-18C at China Lake scored a direct hit against a QF-4 target during the first of five scheduled operational assessment tests. In under than two-years of flight-testing the missile achieved 18 successes in 19-guided flights and a total of 37 successes in 39 launches.

On 1 May 2002 Raytheon delivered the first production AIM-9X and the USAF declared IOC (Initial Operational Capability) in November 2003 on F-15's at Elmendorf AFB, Alaska with the USN/USMC declaring IOC at Iwakuni, Japan in February 2004. In USN service the missile equips the F/A-18E/F Super Hornet, F/A-18C/D Hornet while the USAF employs the weapon on primarily on its F-22 and F-15 fighter; both services will field the weapon on the F-35 Lightning II.

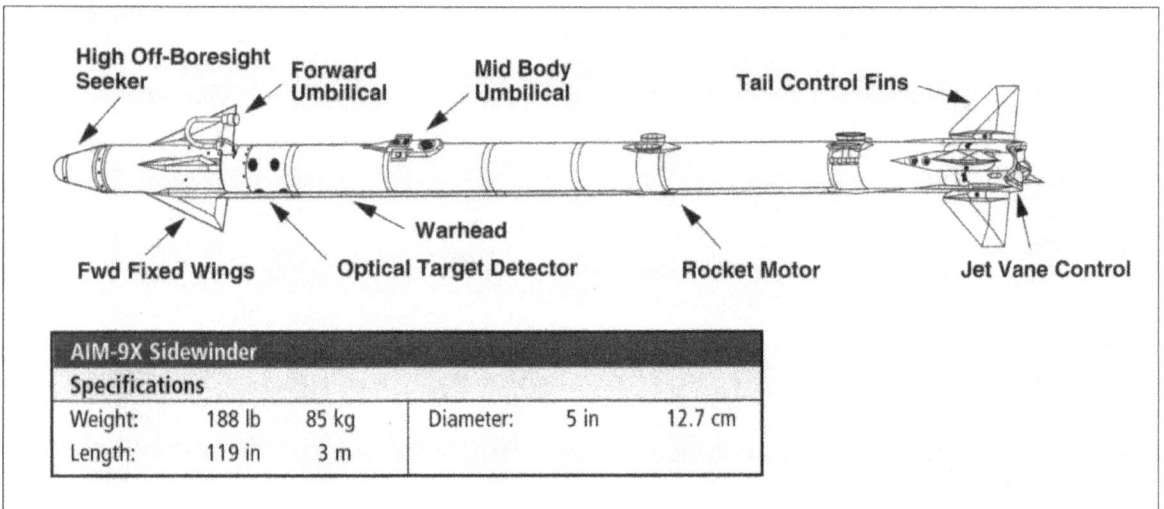

AIM-9X Sidewinder					
Specifications					
Weight:	188 lb	85 kg	Diameter:	5 in	12.7 cm
Length:	119 in	3 m			

Top: Raytheon's Air Defence team for the early part of the 21st Century is the AIM-120 for medium range air-to-air engagements and the AIM-9X for shorter-range engagements. Author **Above: This diagram shows the layout of the AIM-9X.** Raytheon

In March 2004 an AIM-9X was fired for the first time in an operational environment from a US service fighter from the USAF 83rd FWS (Fighter Weapons Squadron) aircraft rather than a test aircraft. This was the 67th AIM-9X launch. On 17 May 2004, the AIM-9X was approved for full-rate production.

The AIM-9X has been ordered by a number of export customers as well as the US. Further developments planned or in development include extending the missiles range as well as introducing software upgrades, allowing the weapon to be used in the air to surface role. A similar program is being conducted in Europe to give the AIM-9L/M an air to surface role by incorporating a laser seeker.

Raytheon offered a three-phase program to meet the RAF FMRAAM requirement. The first phase was the AIM-120B+ (left) followed by the ERAAM (centre) then the definitive FMRAAM (right). Author **Right: Diagram showing the commonality between FMRAAM and AMRAAM.** Raytheon

Raytheon FMRAAM

To meet the UK BVRAAM requirement Raytheon aggressively marketed a three-phased missile program based on the AIM-120 AMRAAM. All three stages would have incorporated the SCAS developed in the AMRAAM P3I phase two program, an improved guidance section, a new electronic safe and arm device and modified digital TDD (Target Detection Device).

In the first stage missile, designated AIM-120B+, the 15.24-cm (6-in) shorter section would be joined to the standard rocket motor of the AIM-120B. An empty structural plug would be inserted between the two parts, therefore, allowing the original length of the missile to be retained. Raytheon claimed that the AIM-120B+ would have been available by 2004, had it been ordered.

The next phase; the ERAAM (Extended Range Air to Air Missile) would have the same front section of the AIM-120B+ joined to a duel-pulse rocket motor 27.94-cm (11-in) longer than the standard AIM-120B rocket motor. This would utilise the extra volume created in the AIM-120B+

and the 5-in SCAS developed during the AMRAAM P3I program. Raytheon claimed that this would have provided around 80% of the capability of the FMRAAM at 50% of the cost.

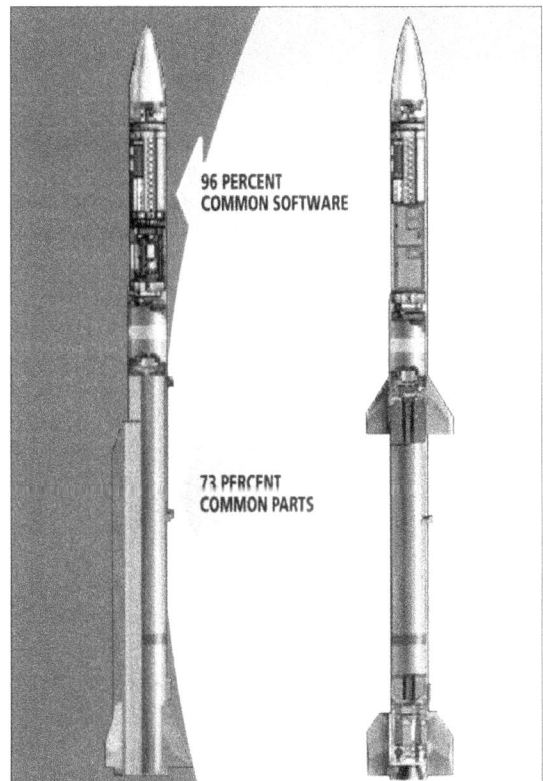

96 PERCENT COMMON SOFTWARE

73 PERCENT COMMON PARTS

Top: A Hughes FMRAAM model seen at Farnborough in September 1996. Above: Following its purchase of Hughes Missiles division, the FMRAAM took on the Raytheon label. Author

On 15 September 1999, Raytheon announced it would offer the ERAAM+ (Extended Range Air-to-Air Missile plus) to develop the next generation BVRAAM. ERAAM+, a hitherto unseen missile variant, would incorporate the latest technology available from the AMRAAM P3I Phase 3 programme.

The third phase of the Raytheon approach would have been the full standard FMRAAM, which would have had the forward section of the ERAAM combined with a rocket boost motor and a liquid-fuel ramjet sustainer, which would be fed by two large air intakes with a fuel tank mounted ahead of the missiles rocket motor. Had it been ordered Raytheon claimed the FMRAAM would have been available from 2007 and estimated that it would provide a 200-300% increase in 'no escape zone' over existing missiles.

The UK was politically pressured from both sides of the Atlantic over the selection of a new missile. The US offered joint development and large production shares of the larger US market. The US government stated that if the UK selected the Raytheon bid for SR(A)1239 then the USAF would adopt the weapon for the USAF Lockheed Martin Boeing F-22 Raptor and the USN Boeing F/A-18 Hornet and Super Hornet. European manufacturers pointed out that the European Meteor was essential to Europe retaining control over export of its combat aircraft, which if armed with US missiles would be subject to US Congressional approval for export sales. This argument won through when the UK MoD selected the Meteor proposal in June 2000.

The US has no current plans to field a missile in the FMRAAM class, although this may change as the AMRAAM becomes progressively outclassed by weapons like the European Meteor and Russian R-77M. In the mid-1990's limited details emerged of a DRM (Duel Role Missile), which could form the basis of an AMRAAM replacement towards the end of the decade.

Chapter Four

Russian Air-to-Air Missiles

A German Luftwaffe MiG-29 launches an R-27 AA'10 'Alamo' medium range air to air missile. Germany inherited the MiG-29 and associated missiles when West and East Germany unified in the early 1990's. US DoD

The first practical Russian guided air-to-air missile was the K-5 AA-1 'Alkali', which was a radio controlled air interception missile that entered service with the air arms of the Soviet Union in 1957. Like early western air-to-air missiles, this weapon was extremely limited in capability, with a range of only 2-6-km depending on launch parameters. In the following four and a half decades Russian missile technology improved dramatically with a number of highly capable SARH, active-radar homing and IR homing missiles now either in service or under development.

R-3 AA-2 'Atoll'

The R-3 AA-2 'Atoll' was developed by the Soviet Union apparently aided by technology gained from the acquisition of an US AIM-9B

Sidewinder IR guided air-to-air missile from China. The AIM-9B was apparently acquired when the missile lodged in a PLAAF MiG-17 fighter during an air combat battle with RCAF (Republic of China Air Force) North American F-86 Sabres over the Taiwan Strait in September 1958.

The R-3 emerged in a similar configuration to the AIM-9B, although technology advances were incorporated into the design with series production beginning in 1960. The first mass produced variant was the R-3S (K-13A), which entered production in 1962.

In 1961, the Soviet Union embarked upon development of the R-3R (K-13R) SARH variant of the missile, which was optimised for high-altitude operations. This variant entered service in 1966.

In the late 1960's, Vympel commenced work on the R-3M (K-13M), a much-improved variant, which received service certification. This variant featured a cooled homing head, a more powerful warhead and incorporated a radio instead of optical closing-in-igniters. A later variant, the R-3M1, was fitted with a modified steering apparatus.

China produced the missile as the PL-2 and improved variants designated PL-3 and PL-5. Romania licence produced the R-3 under the A-91 designation.

R-3 AA-2 'Atoll'

Propulsion: solid rocket motor
Speed: Mach 2.5
Range: 14,000-m

Top: This MiG-25 'Foxbat is armed with two each of R-40R and R-40T AA-6 'Acrid' air-to-air missiles. The SARH and infrared homing R-40 missiles were normally ripple fired with an infrared homing R-40T launched first, followed by the R-40R SARH variant. This ensured that the infrared-guided R-40T did not engage the rocket motor of the R-40R. Above: This Soviet MiG-23 'Flogger' is armed with R-23R AA-7 'Apex' SARH and R-60M AA-8 'Aphid' infrared homing air-to-air missiles.

R-40 AA-6 'Acrid'

In the early 1960's, Soviet fighter aircraft designers were studying high Mach, high altitude interceptors designed to protect the Soviet Unions vast border areas from attack by mass fleets of US long-range bombers armed with nuclear weapons. To arm service variants of these aircraft, Soviet missile designers developed a new generation of air-to-air missiles including the R-40 (K-40), which was developed in both SARH (R-40R) and IR homing (R-40T) variants.

The R-40 missiles were integrated onto the Mikoyan (later RAC MiG) MiG-25 'Foxbat' interceptors equipped with the Smerch-A fire control radar system. This aircraft/missile combination was compromised following the defection of a Soviet pilot with a MiG-25P to Japan on 6 September 1976. Although the MiG-25 was quickly returned to the Soviet Union, Western intelligence had swarmed all over the aircraft prior to its departure gleaning ever ounce of information on the aircraft and its systems. This led to the Soviet introduction of the improved MiG-25PD equipped with the Saphir-25 radar system. New variants of the R-40 missile were developed for the MiG-25PD, designated R-40DR (SARH) and R-40DT) IR guided variants. Later upgrades to these missiles were introduced as the R-40DR1 and the R-40DT1.

Vympel developed the R-40D1 variants after the Molniya Design Company withdrew from air to air missile design work. Production of the R-40 ended in 1991 and the weapon still remains part of the active inventory for the MiG-25 'Foxbat' and MiG-31 'Foxhound' fighters.

R-40 AA-6 'Acrid'

Propulsion: solid rocket motor
Length: 6.2-m
Diameter: 355-mm
Fin-span: 1.8-m
Launch weight: 475-kg
Speed: Mach 4.5
Warhead: 70-kg high explosive fragmentation
Fuse: radar and active laser
Guidance: R-40R command, inertial and semi-active radar; R-40T command inertial and IR

R-23 AA-7 'Apex'

As development of the R-40 progressed for the MiG-25 in the mid-1960's Vympel commenced development of the R-23 (K-23) AA-7 'Apex' to arm the MiG-23 tactical fighter. The R-23 became operational with the MiG-23M variant in 1973.

An R-60E AA-8 'Aphid' short-range air to air missile is shown carried on the port outer wing station of an upgraded Romanian Aerostar MiG-21 Lancer. Author

Like some previous Soviet and later Russian air-to-air missiles the R-23 was developed in both SARH (R-23R) and IR guided (R-23T) variants. Modified variants designated R-24R (SARH) and R-24T (IR guided) were developed, entering service with the MiG-23M2 and MiG-23MLD fighters. The R-24 introduced a number of improvements over the R-23, including a 50-km range for the R-24R compared with 35-km for the R-23R.

The R-23 was produced in the Soviet Union and licence produced in Romania as the A-911.

R-23 AA-7 'Apex'

Propulsion: two-stage solid rocket motor
Length: 4.5-m (R-23R); 4.2-m (R-23T)
Diameter: 0.22-m
Wingspan: 1.0-m
Launch weight: 320-kg
Speed: Mach 3 (3420-km/h)
Range: 35-km (R-23R), 50-km (R-24R); 15-km (R-23T)
Warhead: 40-kg high explosive
Fuse: proximity
Guidance: active radar homing (R-23R and R-24R), passive IR homing (R-23T and R-24T)

Molniya (Vympel) R-60 AA-8 'Aphid'

The Molniya Design Bureau began development of the R-60 (K-16) AA-8 'Aphid' in the late 1960's. The initial R-60 development program was relatively short and series production began in 1973.

This MiG-21 is carrying variants of the R-2 AA-2 'Atoll' infra-red guided air to air missile. The R-2 emerged similar in configuration to the AIM-9B, although technology advances were incorporated into the design with series production beginning in 1960.

The R-60 was designed as a very small IR guided air-to-air missile to arm a number of tactical combat and interceptor aircraft. The missiles small size and lightweight came at a cost. Compared to other IR guided air-to-air missiles of its era, the R-60 had a relatively short-range and lacked punch, with a warhead weighing only 3.5-kg.

Now retired, the Soviet (later Russian) Navy relied on the Yakovlev Yak-38 'Forger' for fleet air defence until the early 1990's. The Yak-38 was a short-take off and vertical landing fighter armed with the obsolete AA-2 'Atoll' or the newer, but extremely short-range, AA-8 'Aphid' seen on the wing stations of this pair of 'Forgers' aboard a Soviet *Kiev* class aircraft carrier. US DoD

Russian MiG-31 'Foxhound' heavy interceptor armed with R-33 AA-9 'Amos' long-range SARH air to air missiles. The MiG-31 was the only aircraft to be armed with the R-33. US DoD

A later development, the R-60M featured all-aspect engagement capability and this weapon was exported to a number of nations as the R-60MK.

Some operators have used the R-60 operationally. It is now clear that an Iraqi Air Force MiG-25 'Foxbat' E fighter engaged and shot down a USN F/A-18C Hornet strike fighter on 17 January 1991 – the first day of the Gulf War. It remains unclear what type of weapon was used, but this may have been an R-60, which armed the MiG-25, as well as the IR and SARH variants of the AA-6 'Acrid'. In December 2002, the Pentagon revealed that an Iraqi MiG-25E intercepted and shot down a USAF

General Atomics RQ-1A Predator UAV (Uninhabited Air Vehicle), although Iraq claimed that ground based air defences had shot down the Predator. If it was an air interception then again it is not known what missile type was used.

The R-60 has been integrated with a number of former Soviet and Russian aircraft including the MiG-21 'Fishbed', MiG-29 'Fulcrum', MiG-23/27, MiG-25, MiG-31, Su-17/20/22 'Fitter', Su-24 'Fencer' and Su-25 ground attack aircraft.

Molniya R-60 AA-8 'Aphid'
Propulsion: solid propellant rocket motor
Speed: Mach 2+
Range: 3-10-km depending on variant
Warhead: 3.5-kg high explosive
Fuse: active-radar and active-laser in R-60M
Guidance: all-aspect passive infrared R-60M

Vympel R-33 AA-9 'Amos'

Russia developed the Vympel R-33 AA-9 'Amos' long-range air-to-air missile as the primary armament for the Mikoyan MiG-31 Foxhound long-range heavy interceptor. The MiG-31/AA-9 combination was developed to defend Russians vast northern borders from attack by US Strategic bombers and to counter the US Mach 3 high altitude Lockheed SR-71 Blackbird strategic reconnaissance aircraft. Despite being designed as a large weapon principally to intercept large bombers and reconnaissance aircraft, the R-33 can operate in all-weather day or night and can engage targets ranging from strategic bombers, low flying cruise missiles to slow moving helicopters in a

Compared to medium-range air to air missiles like the R-27 AA-10 'Alamo' alongside it, the Vympel R-33 AA-9 'Amos' at forefront of picture is a large bulky weapon primarily designed to intercept large heavy strategic bombers like the Boeing B-52H Stratofortress, bringing them down with a powerful 45-kg warhead. Author

dense ECCM environment. In the end phase of the engagement, either an active radar proximity or impact fuse detonates the high explosive fragmentation warhead. The R-33 has a top speed of Mach 4.5 and a claimed engagement range of around 100-km.

The aerodynamic layout of the R-33 includes a cruciform configuration with low-aspect ratio lifting surfaces and two folding control surfaces to allow the missile to be carried semi-recessed beneath the MiG-31 fuselage. Manoeuvring control comes via four differential fins. Guidance is conducted by the semi-active homing on-board radar which locks-on-to the target aircraft in the terminal phase of the flight.

The R-33 was the first Soviet air-to-air missile to be designed with a digital on-board computer system as opposed to analogue. The only airborne platform to operate with the R-33 is the RAC-MiG MiG-31 'Foxhound' long-range interceptor operated by the Russian Air Forces.

There has long been speculation that the R-33 was developed from US AIM-54 Phoenix technology claimed to have been gained by the Soviet Union from Iran. However, development of

the R-33 commenced long before the Islamic revolution overthrew the Shah of Iran in 1979. The basic concept of the R-33 would probably have been set before any possible technology transfer if any took place.

Vympel R-33 AA-9 'Amos'

Propulsion: solid rocket motor
Length: 4.15-m
Diameter: 380-mm
Wingspan: 1.18-m
Speed: Mach 4.5
Range: thought to be around 100-140-km
Warhead: 47-kg high explosive fragmentation
Fuse: active radar
Guidance: inertial, mid-course corrections and semi-active radar homing in terminal phase
Deployed: mid-1980's, on MiG-31 'Foxhound'

Vympel R-27 AA-10 'Alamo'

Both IR and SARH variants of the BVR Vympel R-27 AA-10 'Alamo' air to air missile were developed as the primary armament for the MiG-29 'Fulcrum' and Su-27 'Flanker' B fighters which entered service with the air forces of the Soviet Union in the mid1980's. The R-27 superseded the R-23 AA-7 'Apex' which had armed previous generation fighters like the Su-15 'Flagon' and MiG-23 'Flogger', although this weapon remained the standard weapon for those aircraft.

The R-27 was developed in semi-active radar homing, active-radar homing and IR homing variants. The R-27R SARH variant is shown. Author

The R-27 adopted a cylindrical aerodynamic body with large axially symmetric cruciform control surfaces known as the 'butterfly' configuration allowing the same control surfaces for yaw and pitch control and roll stabilisation.

This RAC-MiG MiG-29K development aircraft is armed with an R-27RE SARH AA-10 on the starboard inboard wing station and an R-73M IR guided missile on the starboard outboard wing station. RAC

A whole family of R-27 missiles variants has been produced, including the R-27R AA-10 'Alamo' A with SARH guidance and the R-27T

'Alamo' B with infrared guidance. Longer range-variants of these missiles were also developed designated R-27RE for the SARH variant and R-27TE for the infrared-guided variant. These missiles designated 'Alamo' C and 'Alamo' D respectively are fitted with a boost sustain motor to extend the range. The longer-range variants are slightly longer than the 'Alamo' A and B variants. Further variants were also developed during the 1990's to increase the weapons capability against targets flying at extreme low altitude such as cruise missiles. Vympel also developed variant featuring active radar homing, although it is unclear, whether this variant entered service in Russia.

The R-27 missile variants still have a long service career ahead of them on a number of Russian aircraft including all MiG-29 and Su-27/30/33/34/35 variants. The Su-34 'Fullback' pre-series production aircraft are regularly seen carrying both the SARH and IR guided variants of this weapon; therefore, it has to be assumed that this weapon will form part of the air-to-air armoury of the new strike aircraft. The standard Su-27 'Flanker' B fighter can carry R-27's on the intermediate wing stations and on the engine intake trunk stations. Although the weapon can be carried on other stations, for instance in tandem on the fuselage centreline stations, this is unlikely while carrying an operational combat load on aircraft tasked with an air to surface mission.

Top: The R-27R and R-27E have been supplied to China as part of the contracts for Su-27/30 fighters. These missile mock-ups are displayed alongside Su-30MKK 502. **Above:** An R-27E SARH AA-10 sits alongside an AS-17 air to surface missile at Farnborough in 1996. Author

While the R-27 has mostly been associated with aircraft developed primarily for air combat, the upgraded Sukhoi Su-25TM 'Frogfoot' ground attack aircraft is capable of operating the R-27R and its successor, the R-77. The Su-25TM's air-to-air prowess comes courtesy of a Kopyo radar system installed in a pod under the fuselage. This allows the aircraft to launch and guide both SARH and active radar homing air-to-air missiles. This model of a Su-25 is shown carrying R-27R SARH and R-77 active-radar homing missiles on wing stations. Author

The R-27R and R-27T have both been widely exported to MiG-29 and Su-27/30 operators. Some export nations like India operates the R-27 variants on both MiG and Sukhoi fighter aircraft.

The R-27 has been used operationally in a few airborne confrontations during the turmoil that followed the break-up of the former Soviet Union. A number of former Soviet Republics operate MiG-29, Su-27 or both types of fighters armed with R-27's. The R-27 has also been used in air combat between Ethiopian and Eritrea Su-27's and MiG-29's, with the Su-27's shooting down a number of MiG-29's.

Vympel R-27 AA-10 'Alamo'

Propulsion: solid propellant rocket motor
Length: 3.70-m
Diameter: 230-mm
Fin span: 0.77-m
Speed: Mach 4
Range: there are a number of differing claims for range for the R-27 variants: From 2-80-km (R-27R), out to 70-km (R-27T), claims of a range of 130-km (R-27RE) and 120-km (R-27TE).
Warhead: 39-kg expanding rod
Guidance: R-27R and R-27RE (SARH), R-27T and R-27TE (passive IR)

This MiG-29K naval fighter is carrying R-27 missiles on the inboard wing stations. RAC-MiG

An underside view of an Su-27 'Flanker' in the 1980's showing the fuselage and wing station carriage of SARH and IR homing variants of the R-27R AA-10 'Alamo' medium-range air-to-air missiles US DoD

Vympel R-73 AA-11 'Archer'

While Russian fighters carry the IR guided variant of the AA-10, this weapon is complemented by the shorter range, but highly agile R-73 AA-11 'Archer' IR guided missile. When it entered service, the R-73 was probably the most advanced short-range air to air missile in the world and was a generation ahead of the latest variants of the Hughes (later Raytheon) AIM-9L Sidewinder or Matra Magic 2 short-range IR guided air to air missiles arming NATO fighters. Only in the first decade of the 21st century is NATO beginning to field comparable systems.

The R-73 was developed to supersede the R-13M AA-2 'Atoll' and the R-60M AA-8 'Aphid' short-range IR guided air-to-air missiles fielded on previous generation Soviet tactical fighters. The AA-11 was developed with all-aspect engagement capability and high agility as a design driver, augmented by the ability of the pilot of the Su-27 or MiG-29 fighters to cue the weapon to targets at up to 60-degree off-boresight via a HMS (Helmet Mounted Sight). The missiles high manoeuvrability is achieved by incorporation of a number of elements including four forward control fins, elevators attached to the rear fins, which are fixed, and deflectors vanes positioned in the nozzle of the rocket engine. Small strakes are fitted ahead of the forward canard-control surfaces. The R-73 uses a combination of air-gas-dynamic control to realise its exceptional manoeuvrability, which allows the missile to engage high agility fighter aircraft, and missiles conducting evasive manoeuvres of up to 12-g. In 1997, Vympel unveiled a variant of the R-73 featuring an active laser fuse, which, it was claimed, had already entered service with the Russian Air Forces at that time.

Two main variants of the basic R-73 were developed; the first, the R-73A has a claimed range of 30-km (19-miles) and the second, the R-73M has a claimed range of 40-km (25-miles). These range figures may be exaggerated; however, the R-73 variants probably have a longer reach than most western equivalents such as the many AIM-9 Sidewinder variants including the latest generation AIM-9X now entering service on USAF and USN fighters.

Above: The R-73 IR homing air to air missile can be integrated with a number of aircraft types ranging from advanced fighter aircraft equipped with advanced sensors to more modestly equipped advanced light attack aircraft. This R-73M is shown carried on a wingtip launch rail of a Russian RAC-MiG MiG-AT advanced trainer masquerading as a light ground attack aircraft. Top: the latest variants of the R-73 in service with the Russian Air Force feature an active-laser fuse. Author

An R-73 mock-up missile is carried on the outboard wing station of a Romanian Air Force Aerostar MiG-21 Lancer, with an AIM-9 Sidewinder inboard. Author

The R-73A became operational on Soviet fighters in 1984 and the weapon has been exported, particularly to MiG-29 and Su-27/30 export customers such as China and India.

Four R-73's can be carried by the Su-27 'Flanker' B; two on the wingtip stations and two on the outboard under wing stations and this would probably be a maximum load for the Su-34. However, the Su-34 would probably carry only two R-73's during operations in a high threat environment when ECM pods would probably be carried on the wingtip stations.

The R-73 is operated by a number of air forces operating MiG-29 and Su 27/30 fighters and is also integrated on a number of other platforms including all members of the Su-27/30/33/34/35 'Flanker' family of combat aircraft. Upgrades of other tactical fighters including MiG-21, Su-24 'Fencer' and Su-25 'Frogfoot' ground attack aircraft have also been integrated with the R-73.

In 1997, Vympel unveiled the K-74M3 (R-74M3), which was an improved variant of the R-73 featuring a more capable IR seeker head, but retaining the same basic airframe as the R-73.

Vympel apparently commenced design work on the K-74 in the mid-1980's as a follow on to the R-73, which was then entering service. The K-74M3 seeker introduced a maximum off-boresight angle of 60 degrees compared with the 40-degrees off-boresight angle of the R-73M. According to Vympel figures, the introduction of the Arsenal Design Bureau IR seeker head increased maximum engagement range by around 30% compared with the basic R-73.

The combination of agile airframe and advanced capabilities of the R-73 slaved to the helmet mounted sight gives the MiG-29 and its larger rival, the Su-27/30 great potential in the close combat arena only now being rivalled by western aircraft. RAC-MiG

A pair of German Luftwaffe MiG-29 'Fulcrum' fighters flies a training mission with USAF F-16C fighters. The MiG-29's are both carrying R-73 training rounds. US DoD

During development of the K-74M3 Vympel studied alternative thrust vector arrangements including the twin inceptor used in the R-73, a four-inceptor design and an all-moving nozzle and four independent inceptors. The twin inceptor design of the R-73 eventually won through, considerably reducing development time and cost. As a cheaper alternative for the home and domestic market, Vympel offered customers an upgrade for existing R-73 missile stocks, which would be fitted with the new IR seeker head of the K-74.

Vympel R-73 AA-11 'Archer'

Propulsion: solid propellant rocket motor
Length: 2.9-m
Diameter: 170-mm
Fin span: 0.51-m
Launch weight: 105-kg (R-73M1) and 115-kg (R-73M2)
Speed: Mach 2.5
Range: 40-km (R-73M2)
Warhead: 17.4-kg high explosive expanding rod
Guidance: all-aspect passive infrared

Vympel R-77 AA-12 'Adder'

Despite the emergence in the early 1990's of the Vympel RVV-AE R-77, development of which commenced in 1982, this weapon was not hurriedly adopted into Russian Air Force service due to insufficient funding following that country's economic woes after the break-up of the former Soviet Union. The weapon is, however, thought to have been in limited service in Russia since 1994. More recently it has been integrated onto systems undergoing limited upgrades to keep them viable until a new fighter developed under the T-50 program, becomes available, perhaps towards the end of the decade. The weapon has also been exported to a number of customers including India and Malaysia and China has also apparently purchased small numbers for its Sukhoi Su-30MKK/MK2/3 and possibly Su-27 fighters.

The R-77 has narrow-span wings of rectangular shape and four lattice control surfaces at the rear. These are of a similar configuration to the control surfaces used on the much larger SS-21 'Scarab' and SS-25 'Spider' ballistic missiles. Among the benefits of this type of control surface is reduced flow-separation at high angle of attack.

The R-77 guidance system includes inertial in the initial phase with mid-course updates via a fighter to missile data link for long-range engagements and active radar homing in the terminal phase of the engagement. The missiles on-board active-radar has an apparent acquisition-range of around 20-km.

The RVV-AE R-77 AA-12 'Adder' is roughly comparable in capability to the US Raytheon AIM-120 AMRAAM. The R-77, however, is thought to have longer range and probably has a higher speed. Author

Like the US Raytheon AIM-120 AMRAAM and the European MBDA MICA EM, the R-77 is a fire-and-forget missile with a multi-stage guidance system, including active radar homing for the terminal phase. The standard R-77 has a superior range to that of the US AIM-120, with unconfirmed reports indicating a range of 100-km (54-nm) at altitude, although this may be an exaggeration. The missile has a laser fuse and can engage 12-g manoeuvring targets. Vympel claims that while the R-77 is heavier than the US AIM-120 AMRAAM and European MICA EM, the R-77 has a longer range and better performance when engaging manoeuvring targets than both of its western rivals. Vympel also claims that the R-77 can be used in a 'self defence' mode to intercept missiles launched at the parent aircraft.

The R-77 has been integrated with a number of Russian combat aircraft including MiG-29 and Su-27 variants, Su-30/33/34/35, upgraded MiG-21 'Fishbed' and upgraded MiG-23 'Flogger' fighters. By summer 1995 the R-77 had been integrated with the upgraded MiG-31M 'Foxhound' Zaslon-M

phased array-radar and would be an armament option for the other variants of this large fighter such as the MiG-31E. The R-77 can also be carried by lightweight combat aircraft like the armed variant of the Yakovlev Yak-130 advanced trainer equipped with radar such as the lightweight Kopyo radar system

Developed primarily as an advanced air to air missile for application to high performance fourth and fifth generation fighter aircraft, the R-77 can also be carried by more primitive aircraft like the lightweight combat aircraft variant of the Russian Yakovlev Yak-130AEM advanced trainer. This Yak-130 model shows the carriage of both R-77 and R-73 missiles. Author

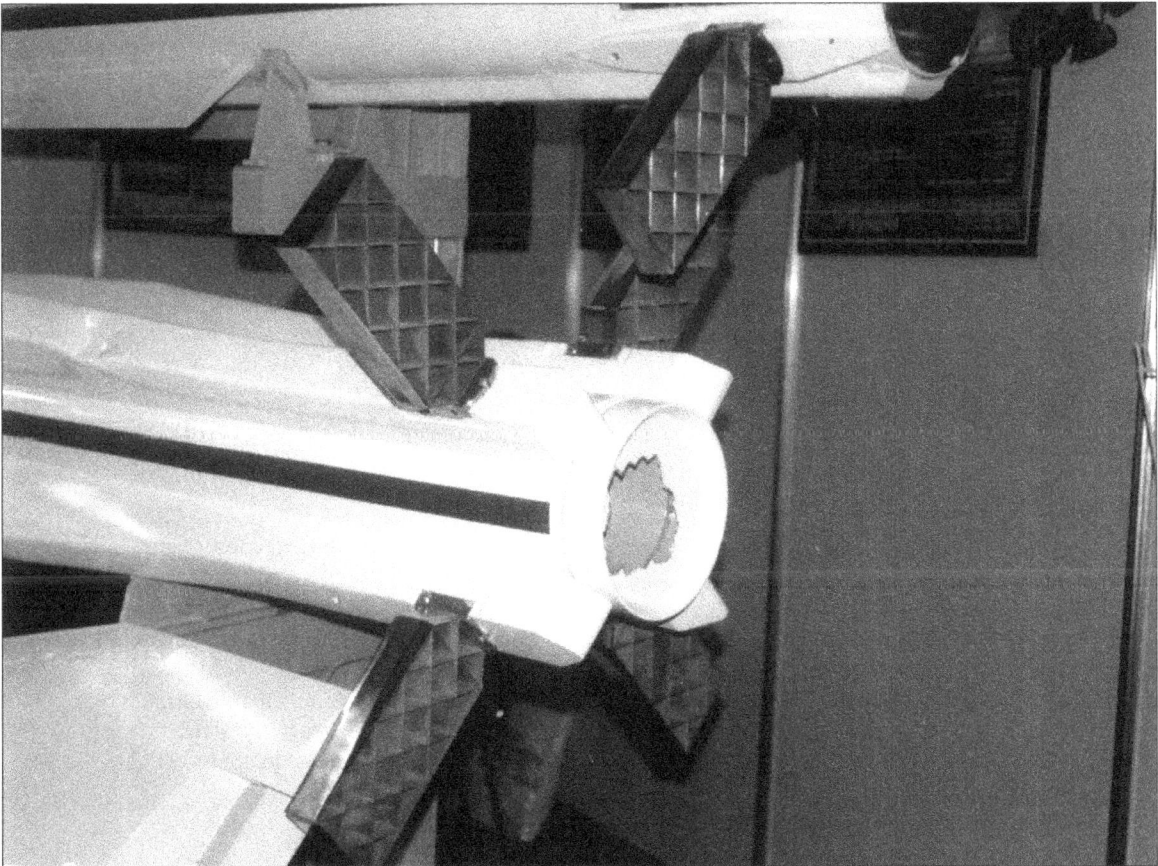

Previous page top: The R-77 is fitted with an Arsenal active-radar seeker head with an antenna diameter of around 200-mm. Previous page bottom: The R-77 has four lattice control surfaces at the rear. Among the benefits of this type of control surface is reduced flow-separation at high angle of attack. Author

Above: Russia is developing a long-range variant of the R-77 known as the R-77M fitted with a ramjet sustainer, which kicks in after rocket motor burnout. The R-77M is shown at bottom of photograph with a standard R-77 above. Author

---

Vympel R-77 AA-12 'Adder'

Propulsion: solid propellant rocket motor
Length: 3.60-m
Diameter: 200-mm
Wingspan: 0.35-m
Launch weight: 175-kg
Speed: Mach 4
Warhead: 30-kg high explosive
Fuse: active-radar
Guidance: inertial, command and active-radar in the terminal phase

---

R-77M

At the 1997 Moscow Air Show, MAKS-97, Vympel released details of the R-77M, which was understood to be fitted with a new improved active-seeker from AGAT and have its maximum range increased to 160-km (86.5-nm) at altitude. The missile is fitted with ramjet sustainers, which takes over when the rocket motor burns out. It is doubtful that the R-77M is near operational service and lack of funding may have delayed the program indefinitely, although Vympel was still marketing the missile at Farnborough 2000, with a mock-up missile displayed alongside the R-77 model. The missile may be offered on the export market to countries such as India and China who may contribute to funding the program as a way to counter future Western air to air missiles like the pan-European Meteor active radar guided air-to-air missile, which has a planned operational range in excess of the current R-77 variants. Development is thought to be continuing, though slowly, and the weapon may be a potential armament for the new advanced fifth generation fighter under development in Russia.

This MiG-35 is carrying R-27 (AA-10), R-77 (AA-12) and R-73 (AA-11) missiles. As well as integrating advanced air to air missiles onto modern aircraft like the MiG-35, Vympel has integrated the R-77 and R-73 onto older aircraft such as the MiG MiG-23 'Flogger' and MiG-21-93 'Fishbed'. RAC-MiG

In January 2000, it was reported that that Russia and China were negotiating plans for a joint Ramjet missile development program to develop an ultra long-range, high-speed air-to-air missile. Russian manufacturer Vympel confirmed in an interview that it was in detailed discussions with China about joint development of a ramjet version of the R-77 missile, which had apparently already completed successful ground tests. The missile and its associated technology are thought to be available for sale or licensing to other nations and Vympel are quoted as saying it could be in service within four years of a contract award.

Powered by an air-breathing ramjet, rather than by today's traditional rocket motor technology, such missiles have considerably extended range, greater speed and exceptional manoeuvrability. Seen as an eventual replacement for the R-77, the ramjet version is expected to equip the Sukhoi Su-27 and Su-30/35 combat aircraft family as well as the fifth generation fighter under development in Russia.

Vympel R-37 (K-37)

In the 1990's, the Vympel K-37 (R-37) was being developed as a successor to the R-33 AA-9 'Amos' long-range air to air missile to arm upgraded MiG-31M interceptors then planned for the Russian Air Forces. However, like so many programs in Russia, the K-37 suffered from an acute funding shortfall in the economic hardship that gripped Russia following the break up of the Soviet Union. In mid-1995, Vympel claimed that it was still some two-years from completing development of the R-37 and around this time the R-37 was undergoing launch trials at the Russian Ahktubinsk test centre in southern Russia.

Vympel information claims that the R-37 would carry a more powerful 60-kg warhead and have longer range than its predecessor, the R-33, which has a warhead weight 45-kg, both missiles apparently utilising the same rocket motor. The increase in range would be gained through the use of more efficient ballistic trajectories when intercepting the target.

The R-37 is capable of intercepting a wide range of targets from high performance fighters to low-flying cruise missiles. Test missiles are thought to have weighed in around 500-kg, although Vympel intended to reduce this to around 450-kg for service weapons. While the R-33 had two folding fins to allow four of the weapons to be carried on under fuselage stations, the R-37 has four folding rear wings. The diameter of the wings of the R-37 has also been reduced compared with the R-33.

This early schematic shows a Western intelligence interpretation of the then Soviet AA-11 'Archer' in the 1980's. US DoD

This Sukhoi Su-35 4ᵗʰ+ generation fighter is carrying R-77 and R-73 missiles. Sukhoi

Vympel has described the R-37 as a statically unstable aerodynamic design, unlike the R-33, which was stable. The R-37, therefore, required an advanced flight-control system, which also endowed the missile with increased manoeuvrability over the R-33. The R-37 would be equipped with an active-radar guidance homing-head, a major improvement over the SARH guidance system used in the R-33.

Once expected to become operational on the Russian Air Force planned fleet of MiG-31M 'Foxbat' heavy interceptors, the status of the R-37 program is unclear. A lack of funding in a Russian economy struggling to recover from the break up of the Soviet Union and the internal turmoil which effected the region since has left scarce funds for many civil and defence programs.

The MiG-31M has all but fell by the wayside, with RAC-MiG now pushing more versatile multi-role variants of the fighter under the MiG-31BM and MiG-31E designations. Even if these variants enter service, it remains unlikely in the current climate that they would be equipped with the R-37, relying on the much shorter range R-77 instead.

Russia is still working on long-range air-to-air missile programs although whether funding will become available to field any such system remains uncertain. In 2000, Agat unveiled a new active-radar seeker, which was aimed at long-range air-to-air and medium-range surface to air missile applications. Previous Agat radar seeker's had a maximum length of around 200-mm such as that installed in the R-77. The new larger seeker designated 9B-1103M is 330-mm and weighs in at some 13-kg (28-lb). According to Agat, the seeker can acquire a 5m2 fighter aircraft sized target at a range of no less than 40-km (21-nm). The new seeker may be aimed at either the R-37 or the Novotor KS-172 AAM-L long-range air-to-air missile, which has also been under development as a possible armament for the Su-35 Advanced Flanker or any possible follow on programs.

The KS-172 was being developed as an ultra-long-range air-to-air missile with a projected effective range of around 400-km (248 miles). This would allow the missile to be used against high value targets such as AWACS (Airborne Warning and Control System) aircraft like the Boeing E-3 Sentry with the reduced need to over fly hostile territory.

The KS-172, which has an active-radar seeker head, would use inertial mid-course guidance corrections before acquiring the target with its own on-board radar in the terminal phase. The missile can be used at altitudes up to 30000-m (98,425-ft) allowing the interception of high flying reconnaissance aircraft.

The KS-172 was being pushed as a possible armament option for the Sukhoi Su-35 Advanced Flanker' and the Russian MFI. With the MFI cancelled and the Su-35 program merely ticking along, the KS-172 has also apparently fallen into obscurity, although it is unclear if the program has been cancelled or may be resurrected to field a possible long-range armament for the Russian LFI fifth generation fighter program under development as the T-50.

Chapter Five

Chinese Air-to-Air Missiles

A Chinese PLNAF (Peoples Liberation Naval Air Force) Shenyang J-8IID fighter is shown carrying a pair of PL-8 short-range air-to-air missiles while intercepting and escorting a USN EP-3 reconnaissance aircraft. USN

PL-2

In the 1960's, China began to develop the PL-2, which was a reversed engineered variant of the Soviet R-3 (K-3) AA-2 'Atoll' short-range IR guided air-to-air missile. Like the AA-2, this was a primitive first generation IR guided missile, production of which commenced in 1964 for the PLAAF (Peoples Liberation Army Air Force) Around this time the improved PL-2A variant entered development with service entry commencing in 1970. In 1978, China began development of the PL-2B, which featured a number of improvements including a more sensitive IR seeker. This variant entered service with the PLAAF in 1981. As well as serving with the PLAAF, the PL-2 variants were widely exported along with Chinese built fighters to several nations including Albania, Bangladesh,

Burma, Iran, Iraq, Pakistan, Sri Lanka, Sudan, Tanzania and Zimbabwe. The missile is integrated with Chinese J-5, J-6, J-7, J-8II and Q-5 fighter and attack aircraft. Further refinements to the design resulted in the PL-3, which was still derived directly from the Soviet R-3

China cleared the PL-8 for export and integrated the weapon onto the Chengdu J-7MG (F-7MG) fighter, which is derivative of the Soviet MiG-21 developed in China. The F-7MG has been produced for the export market. This F-7MG model is shown carrying PL-8 missiles. Author

Currently the most advanced short-range air to air missile produced in China is the IR guided PL-9, which was developed to arm China's new generation of combat aircraft including the J-10 and FC-1 (JF-17). Author

PL-2

Propulsion: solid propellant rocket motor
Length: 2.99-m
Diameter: 0.13-m
Wingspan: 0.53-m
Launch weight: 75-81-kg depending on variant
Speed: Mach 2.5
Range: 3-km
Warhead: 11.3-kg blast fragmentation
Guidance: IR

PL-5

The PL-5 was developed by the Luoyang Electro-Optics Technology Development Centre (EOTDC) for service with a variety of Chinese built fighter and attack aircraft. Obsolete Chinese fighter aircraft such as the J-7B, Q-5 and JH-7 fighter-bomber are armed with the PL-5. Development of this missile began in April 1964, with both SARH and IR guided variants planned. The basic design of the PL-5 adopted a number of features found on

the PL-2 (Soviet AA-2 'Atoll' copy). The new missile, however, introduced a greater engagement range and improved capability IR seeker head.

The first variant was the PL-5A SARH missile, which was cancelled in the early 1980's after being a decade and a half in development. Development of the PL-5B passive IR guided variant began in the mid-1960's, with the first launches taking place in 1967. However, a protracted development period meant that the missile did not enter PLAAF service until 1986; some two decades after development began. Despite the long-development period, the PL-5B, like the PL-2 before it, lacked all-aspect capability and was, therefore, not able to compete on even terms with Western missiles like the AIM-9L/M or R.550 Magic II.

The prototype Chengdu FC-1 (JF-17) takes-off on its maiden flight carrying dummy PL-9 IR missiles on the wingtip launch stations.

This model of the Chengdu Super 7 (FC-1/JF-17) shows the carriage of PL-9 IR homing missiles on the wingtip stations and PL-11/12 radar homing missiles on the outboard wing stations. Author

Improved off-boresight capability, more sensitive seeker and increased manoeuvrability were introduced with the PL-5C variant, which is the main PL-5 variant in PLAAF and naval aviation service. The PL-5E introduced re-designed control surfaces, increasing g-limits to 40-g. All-aspect engagement capability was introduced with a 4-degree off-boresight capability. Increased tolerance to enemy countermeasures was also introduced. It is unclear if this variant entered PLAAF service.

PL-5

Propulsion: solid propellant rocket motor
Length: 3.13-m (PL-5B/C), 2.89-m (PL-5E)
Diameter: 0.127-m
Wingspan: 0.657-m
Launch weight: 148-kg (PL-5B/C), 83-kg (PL-5E)
Speed: Mach 2.2
Range: 1.3-16-km (PL-5B/C)
Warhead: 6-kg high explosive blast fragmentation
Fuse: proximity
Guidance: rear-on aspect IR (PL-5B/C), all-aspect IR (PL-5E)

PL-7

The PL-7, which is a short-range IR (InfraRed) guided air-to-air missile developed by the Zhuzhou Aerengine Factory (ZAF) in China, has apparently been developed from the French Matra R.550 Magic IR missile. The development program commenced in 1982 with initial production missiles being delivered in 1987.

The PL-7 has been credited with good manoeuvrability, automatic target seeking and acquisition, homing and off-boresight capability. It is thought that the missile is fitted with an indigenous Chinese developed IR guided seeker, which is speculated, to be inferior to the original French seeker in the Magic. Unlike the later Magic II, the PL-7, like the original Magic 1, does not feature all-aspect engagement capability. The missile is thought to be operational with Chengdu F-7M Airguard fighters and A-5/Q-5 Fantan ground attack aircraft delivered to export customers. However, the missile has not been noted on any PLAAF or PLNAF aircraft and is, therefore, not thought to be in operational service with China's armed forces. With the availability of more modern missiles it would seem doubtful that the PLAAF or PLNAF (Peoples Liberation Naval Air Force) are operating with PL-7's.

This almost front-on view of the Super 7 model displayed at Paris in 2001 shows the wingspan of the PL-11/12 missiles on the outboard wing stations. This weapon bears a resemblance to the MBDA Aspide SARH air-to-air missile from, which the Chinese programs apparently drew technology. Author

---

PL-7

Propulsion: solid propellant rocket motor
Length: 2.75-m
Diameter: 0.157-m
Wingspan: 0.66-m
Launch Weight: 90-kg
Warhead: 12.5-kg high explosive
Speed: Mach 2.5
Range: 0-4-14-km (some sources claim a maximum range of 7-km)
g-limits: 35-g
Guidance: tail-aspect IR seeker

PL-8

Along with the PL-5, the PL-8 is the main air-to-air missile in service with the PLAAF and Chinese naval aviation. J-7E, J-8B (PLAAF), J-8D (naval) and the fourth generation J-10 normally carry the weapon. The PL-8 was, which is basically a Chinese copy of the Israeli Rafael Python 3 IR guided missile, was developed by the Louyang

China Academy of Air to Air Missile. In 1982, China and Israel apparently signed a co-operation agreement for missile production and technology transfer to China. The PL-8, the first of which were apparently delivered from 1988, introduced an all-aspect IR guided air-to-air capability to the PLAAF when it entered service in the early 1990's. The extremely capable missile can be integrated with a HMSS (Helmet Mounted Sighting System) giving greater operational capability.

PL-8

Propulsion: solid-propellant rocket motor
Length: 2.99-m
Diameter: 0.16-m
Wingspan: 0.81-m
Launch weight: 115-kg
Warhead: 10-kg high explosive
Speed: Mach 2
Range: 15-km
g-limits: 35-g
Guidance: all-aspect IR seeker (later variants can be cued by a HMSS)

PL-9

The PL-9 is an IR guided short-range air-to-air missile developed by the China Academy of Air to

Air Missile/Luoyang Xi'an Aeroengine Factory (XAF) for service with the PLAAF and export customers, with initial batch missiles apparently entering service with the PLAAF in the early 1990's. Development of the 3rd generation PL-9 commenced in 1986, with initial production commencing in 1989, before the missile made its western public debut at the 1991 Paris Air Salon in June 1991.

The IR seeker of the PL-9 features all-aspect capability and the missile is claimed to have better manoeuvrability and all-round performance than the Raytheon AIM-9L/M Sidewinder variants. There are reports, some based on sketchy evidence or even pure speculation, that the PL-9 is based on the Israeli Rafael Python 3. However, externally there is little family resemblance with most if any Python 3 derived technology being internal.

It is thought that the PL-9 has been integrated with a HMSS; either a completely indigenous design or a model copied from the Arsenal HMSS used to cue the Russian Vympel R-73 AA-11 'Archer' IR guided air to air missile. The latter HMSS is claimed to have a 60-degree off-boresight capability, with a 120-degree claimed field-of-fire. Applications for this HMSS are possibly the F-7MG and F-8IIM series of fighters developed for export. The PL-9D is a SAM (Surface to Air Missile) developed from the PL-9C air-to-air missile. This weapon has apparently entered service with PLA (Peoples Liberation Army) ground forces. The PL-9 has been integrated on Chinese J-7 and J-8 fighters and may have been exported to a number of nations including Pakistan and Iran as well as serving with the PLAAF.

PL-9

Length: 2-90-m
Diameter: 0.157-m
Wingspan: 0.65-m
Launch weight: 115-kg
Speed: Mach 2.1
Range: 21-km
g-limits. 35-g
Warhead: 12-kg high explosive
Guidance: all-aspect IR

PL-11

The PL-4 was a claimed Chinese copy of the US AIM-7 Sparrow SARH MRAAM development of which began in 1966 and was apparently completed in 1980. Further development began in

1981 with an initial production batch of missiles possibly delivered in 1984. However, the program was cancelled in 1985.

China's ambitions to develop an indigenous medium-range air-to-air missile remained, however, and in the late 1970's, long before the PL-4 was cancelled, China embarked upon development of the PL-10 SARH MRAAM as the planned primary armament of the Shenyang J-8B interceptor. The PL-10 was developed from the ship/ground launched HQ-6 SAM. The missile entered flight-testing from 1982, with the first launch from a J-8B taking place in 1986. Development of the PL-10 was abandoned in the late 1980's as performance was lower than that required.

Once the PL-10 development was stopped, China embarked upon development of the PL-11, which apparently drew on technology gained from the Italian Aspide, a SARH air-to-air missile, which was armed Italian Lockheed F-104S Starfighters (now retired). An agreement was apparently concluded between Italy and China in the late 1980's for purchase of a small batch of Aspide missiles along with associated technology transfer to China. The co-operation between Italy and China came to a sudden halt when the agreement was terminated following the Tiannamen Square killings in 1989.

The photograph of an Aspide SARH missile being launched from an Italian F-104 Starfighter (top) shows the strong resemblance to the Chinese PL-11/12 (bottom). MBDA and Author

In 1990, China began developing an improved PL-11 designated PL-11A, which also drew on Aspide and HQ-6 technology. Small numbers of PL-11A missiles are thought to have been produced and may have been delivered to the PLAAF as an interim MRAAM for the J-8B pending later deliveries of more advanced weapons. An improved variant of the missile was developed as the PL-11B with an active radar seeker. Technology and experience gained in developing this weapon would be passed onto later programs.

The CATIC Super 7 (FC-1) mock-up is shown carrying SD-10 (PL-11/12) active-radar guided missiles on the under wing stations. PAF

PL-12

While the PL-10 and PL-11/A missile development programs proved a disappointment for China, the quest for an indigenous MRAAM continued with development of the PL-12, which was being developed as the primary armament for the Chengdu J-10 and FC-1 multi-role fourth generation fighters. In 1996, THE china Leihua Electronic Technology Research Institute (CLETRI) revealed a flat-plate antenna for the AMR-1 active-radar seeker, with a gimballed mechanism for use on the PL-11/12. The antenna apparently resembled the Russian AGAT 1103M antenna of the active-radar guided variant of the Vympel R-27.

The PL-12 was publicly revealed at the Zhuhai Airshow China in 2002 as the SD-10, where it was revealed that the weapon had been under development since the 1990's. There is speculation that the PL-12 and SD-10 were different programs, although it may is possible that the PL-12 is the domestic variant for the PLAAF and the SD-10 is the export designation of the missile.

Like the PL-11 before it, the PL-12 is thought to have benefited from technology gained from Italy's Aspide SARH missile, while the AMR-1 seeker which manufacturer information shows is fitted in the SD-10, has apparently been developed from Russian seeker technology. By the early 2000's, the new PL-12/SD-10 is thought to have undergone ground test firings and captive carry trials on a Chinese J-8B fighter.

In the mid-2000's it was considered that the PL-12, being developed by the Leihau Electronic Technology Institute (LETRI), also known as the 607 Institute, could enter service around 2007.

Chapter Six

Rest of the World Air-to-Air Missiles

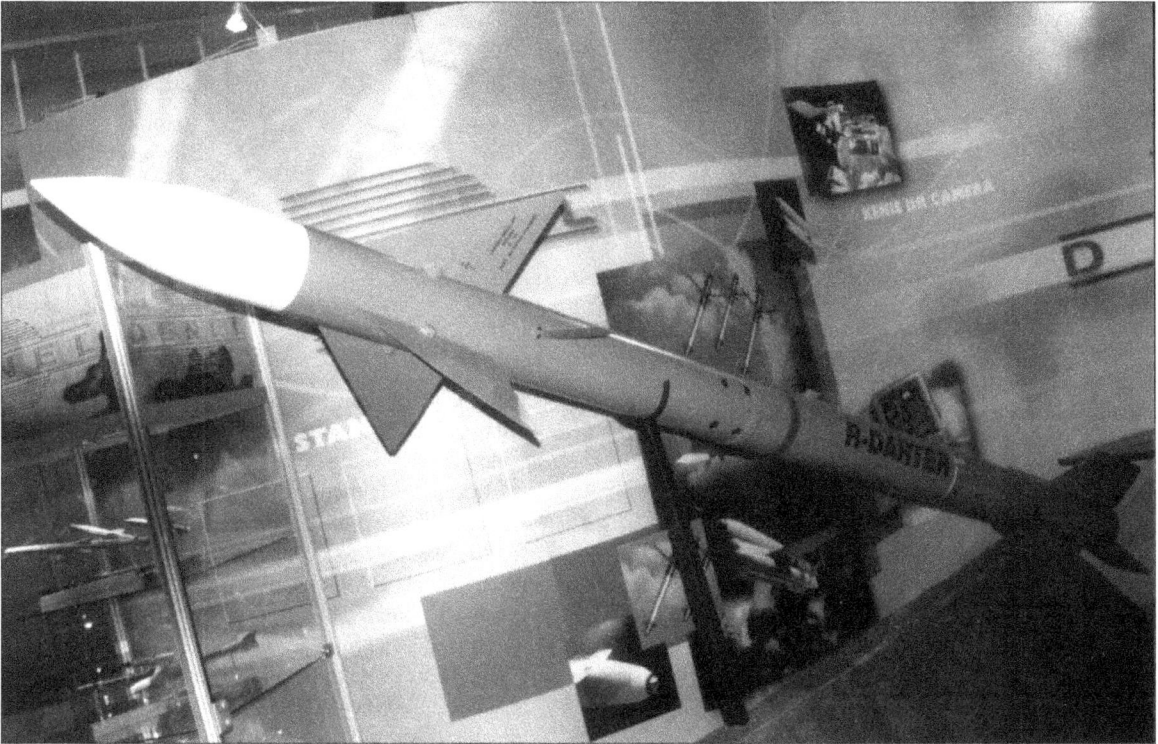

South Africa's R-Darter and Israel's Derby active-radar guided air to air missiles look like identical twins, a result of obvious co-operation between the two nations in development of the missiles. Author

DRDO Astra

Although operating French R.530D and Russian R-27 semi-active radar guided and Russian R-77 AA-12 'Adder' active radar guided air to air missiles for its Mirage 2000H, MiG-29 and Su-30MKI fighters, India has embarked upon a program to field an indigenous active radar guided air to air missile for service with the HAL LCA (Light Combat Aircraft) and other aircraft. The Astra program is headed by India's DRDO (Defence Research and Development Organisation). The feasibility study commenced in the early 2000's, with planned fielding of a weapon system around 2010/12.

Little hard facts of the program are known. A terminal active radar guided seeker head will be included with the ability to receive mid-course targeting updates. The missile will be equipped with an advanced ECCM suite to enable it to operate in a high threat electronic warfare environment, with the ability to jam signals coming from threat radar.

Performance characteristics are thought to be similar to the Russian R-77, while the aerodynamics are thought to be centred around a more slender and longer MBDA R.530D currently operated by Indian Air Force Mirage 2000H fighters. Preliminary information shows that the missile length is 3.7-m, with a body diameter of 178-mm. Launch weight is claimed to be around 154-kg with a 15-kg pre-fragmented warhead with radar proximity and later laser proximity fuses.

DRDO Astra

Length: 3.70-m
Diameter: 178-mm
Weight: 154-kg
Propulsion: solid fuel rocket
Warhead: 15-kg pre-fragmented
Guidance: inertial, with mid-course targeting updates via data-link with active-radar homing in the terminal phase
Range: planned to be 80-100+ km in a head-on engagement

Grainy video still of a Denel U-Darter infrared-guided air to air missile launched from the wingtip launch rail on a South African Dassault Mirage F1 fighter. Denel Dynamics

U-Darter/A-Darter/R-Darter

The Denel Dynamics (formerly Kentron) A-Darter is a fifth generation IIR guided short-range air to air missile system, (co-funded by Brazil), developed for service with the SAAF (South African Air Force) and export as a replacement for the U-Darter, which was a South African IR missile developed from the Matra R.550 Magic 1.

An A-Darter IR guided missile is launched from a ground-test rig during development. Denel Dynamics

The missile is 2980 mm in length, 166 mm wide and weighs 93 kg. A-Darter aerodynamics incorporates features optimised for high agility, including thrust vector control for improved manoeuvrability over legacy missiles, and extended range. The missile features a two-colour thermal imaging seeker with high sensitivity and a multi-mode ECCM suite and lock-on-after launch and memory tracking capability. As well as being directed to the target by the aircraft radar or Helmet Sighing System, the missile can use its autonomous scan feature, which allows the launch aircraft to remain radar silent. A combination of wide-look-angle seeker and a helmet sighting system allow high-off-boresight launches to be conducted.

Flight trials commenced in 2010 and it is planned for the weapon to be operation on SAAF Gripens in 2014, and BAE Systems Hawk 120's later, with a small batch of European IRIS-T missiles purchased as a stop gap.

The Denel Dynamic R-Darter is an all-weather, all-aspect active-radar guided BVR air to air missile designed to engage manoeuvrable targets from short to medium ranges. The system entered service with SAAF Atlas Cheetah C fighters in summer 2000 designated V4, but was retired in 2008 when the Cheetah C's were withdrawn from service. It was expected that Brazil would order the missile, but so far no order has materialised.

This artist impression shows a Denel A-Darter being launched from a South African Air Force Saab JAS 39 Gripen 4th Generation fighter. Denel Dynamics

Mock-up of the Rafael Derby shown on the port side wingtip station of an IAI (Israeli Aircraft Industries) F-16 Ace upgraded F-16 in 2001. Author

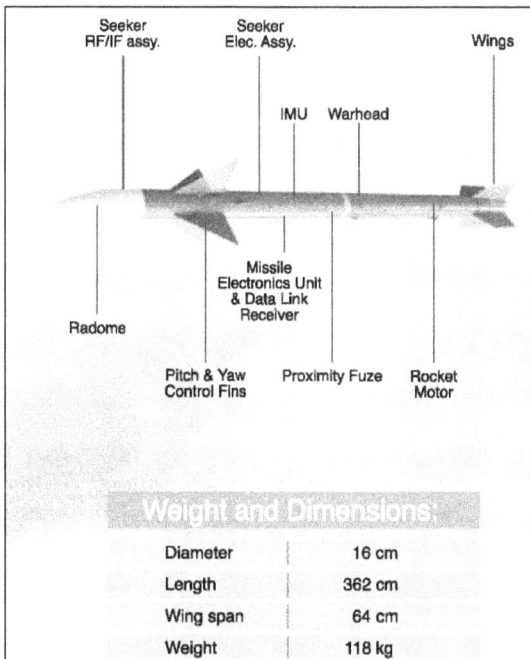

Weight and Dimensions:	
Diameter	16 cm
Length	362 cm
Wing span	64 cm
Weight	118 kg

Derby entered operational service with the IDF in 2003 and it is claimed was used successfully to shoot down a claimed Hezbollah UAV in 2006, although this has not been independently confirmed. Rafael

Rafael Derby

The Rafael Derby, which was publicly revealed in June 2001, was developed as a beyond visual and within visual range active-radar guided air-to-air missile to arm IDF (Israeli Defence Force) fighter aircraft. Derby is claimed to feature outstanding manoeuvrability when used in both medium and short-range engagements. The active-radar seeker has been designed for use in all-weathers and in a dense electronic counter measures environment. The missile can be used in look down/shoot down mode and features a LOBL (Lock-On Before Launch) capability when used for close-range manoeuvring dogfights. Resistance to jamming comes courtesy of an advanced programmable ECCM suite. The Missile has a diameter of 16 cm, a length of 362 cm, and a span of 64 cm and weighs in at 118 kg.

Derby Key Features include Active Radar; All-Weather capability; Look-Down/Shoot-Down; LOAL (Lock-on After Launch) and LOBL (Lock-on Before Launch) Capability; Duel Range (medium/short); Data Link Receiver; Fire and Forget; Large Kinetic Envelope; Advanced ECCM capabilities; High Agility and Light Weight

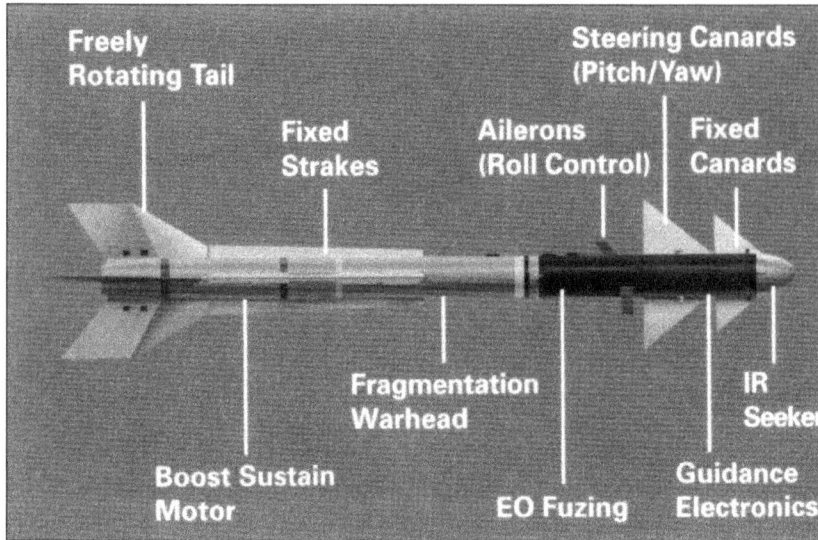

Labels on diagram:
Freely Rotating Tail
Fixed Strakes
Ailerons (Roll Control)
Steering Canards (Pitch/Yaw)
Fixed Canards
Fragmentation Warhead
IR Seeker
Boost Sustain Motor
EO Fuzing
Guidance Electronics

Rafael Python 4. Rafael

A Romanian IAR-109 Swift advanced jet training aircraft is towed at Farnborough carrying a Rafael Python 3 missile on the port side wing station. Author

Rafael Python 3 and Python 4

Israel is one of the world leaders in development and fielding of IR-guided air-to-air missiles such as the Rafael Python 3 third generation IR guided missile designed for shot range combat. The missile has been integrated with a number of aircraft including the F-15, Dassault Mirage series, F-16, F-4 Phantom II, Northrop Grumman F-5 and Israel's Kfir C-2/7 fighters, which were developed from the Mirage III/5 designs.

The Python 3 can be used in head-on and beam engagements and against targets engaging in high g tight turning dogfights. The all-aspect capability allows it to be used in head-on engagements against closing targets. It can be used against targets ranging from high performance fighter

aircraft to low-flying helicopters. The missile, which is credited with a range of around 15-km at high altitudes, can be used by aircraft with relatively unsophisticated on-board systems. The missile can be slaved to the launch aircraft's radar with a 30-40 degrees cone of axis. Once launched the missile closes with the target which is destroyed by an 11-kg warhead.

Python 3 has been used in combat by the Israeli air force and has been credited with shooting down a large number of Syrian Air Force fighter and attack aircraft, particularly during the 1982 air war over the Beka Valley.

Rafael Python 3

Propulsion: solid propellant rocket motor
Length: 3-m (9-ft 10-in)
Diameter: 160-mm (6.3-in)
Wingspan: 0.89-m (2-ft 9.9-in)
Launch weight: 120-kg (264-lb)
Speed: Mach 3.5
Range: from 0.5-km out to around 15-km at altitude
Warhead: 11-kg (24.25-lb) high explosive
Guidance: IR homing head

The Python 4 fourth generation IR guided short-range air-to-air missile, which is regarded as among the most capable short-range missile in service, is operational on a number of Israeli air force fighter aircraft. Designed for instantaneous manoeuvrability off the launch rail, the missiles extreme manoeuvrability is matched by the very advanced IR seeker head, which bestows a high probability of a successful engagement leading to the shoot down of the target aircraft.

The advanced seeker and efficient aerodynamic layout both contribute to the increased "no escape zone" over previous generation IR guided missiles. This zone extends to most of the pilots frontal hemisphere, therefore, allowing the pilot to engage any potential target entering this zone regardless of the targets flight path or evasion manoeuvring tactics. Python 4 features an advanced IRCCM (Infra-Red Counter Counter Measures) suite, which contributes to the missiles capability to operate successfully when encountering enemy countermeasures.

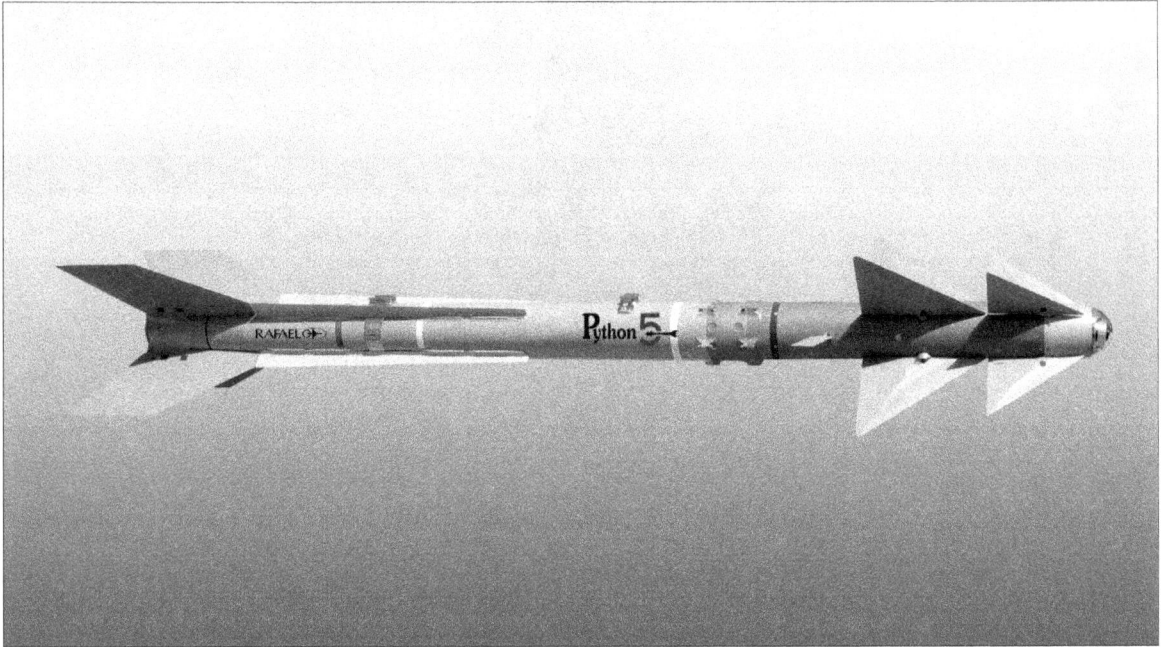

Python 5

Full Sphere IR Missile

Python 5 is a fifth generation air-to-air missile and the latest member of the Python family.

The missile provides the pilot engaging an enemy aircraft with a revolutionary full sphere launch capability. Python 5 can be launched from a very short range to beyond visual range with greater probability of kill and excellent resistance to countermeasures.

Python 5 revolutionary full sphere capability is achieved by the lock-on-after-launch feature combined with excellent acquisition and tracking capability.

The Python 5 missile offers superior war fighting capability to ensure air superiority for pilots in the 21st century.

Python 5 Specifications

Weight	103.6 Kg
Length	3096 mm
Wing Span	640 mm
Diameter	160 mm

Technology Advancement

Python 5 is an innovative combination of state-of-the-art technologies and proven Python 4 components.

The missile incorporates a new dual waveband, high off-boresight imaging seeker, advanced computer architecture, sophisticated IRCCM and flight control algorithms, while maintaining Python 4 unique aerodynamic airframe, powerful rocket motor, warhead and proximity fuze.

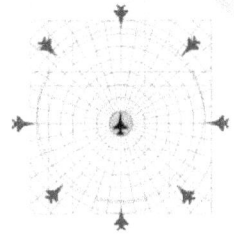

Full sphere launch envelope

Operational Benefits

- Full sphere launch envelope from very short to beyond visual ranges
- Excellent acquisition and tracking performance.
- Target Lock-On-Before and After Launch capability (LOBL & LOAL Modes)
- Excellent resistance to countermeasures
- Greater probability of kill over wider encounter conditions and spectrum of target types.

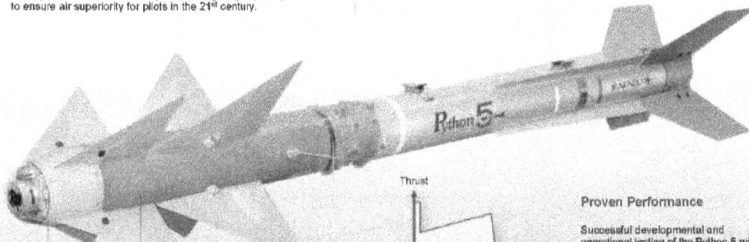

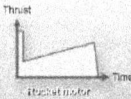

Proven Performance

Successful developmental and operational testing of the Python 5 missile has already been carried out, including extensive captive carry evaluation and homing tests.

Python 5 has demonstrated outstanding target detection and tracking in adverse background and clouds environment.

Python 5 seeker imagery

Imaging seeker

FOG based INS

Electronics unit including signal processing

Thrust / Time — rocket motor

Previous page top: A Python 4 is shown with a Saab Gripen mock-up in 2001. Previous page bottom: A Python 4 is shown on the outboard under wing station of a USAF F-16C Block 50 masquerading as the F-16 Block 60 at Farnborough in 2000. Author

Above: Rafael has developed the Python 5, which retains the same basic layout of the Python 4, but introduces a more capable duel-waveband imaging seeker head and IRCCM (InfraRed Counter Counter Measures). Rafael

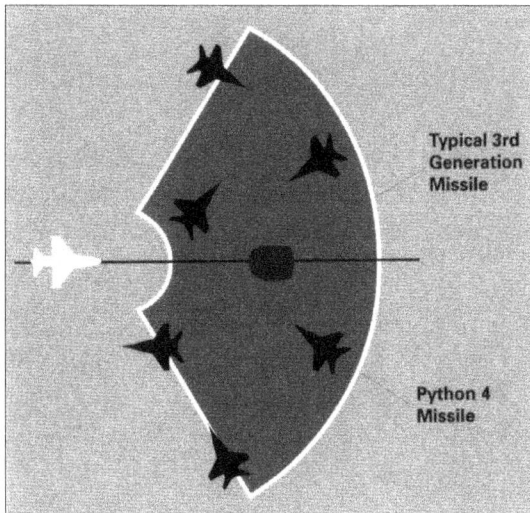

Above: **The Python 4 introduced a number of improvements over its predecessor and other third generation missiles including an increased engagement zone. The Python 4's engagement zone is shown in comparison to the typical 3rd generation missile engagement zone, which is indicated by the dark circular area in the centre.** Rafael

---

Rafael Python 4

Propulsion: solid-propellant rocket motor
Length: 3-m (9-ft 10.1-in)
Diameter: 160-mm (6.3-in)
Wingspan: 0.86-m (2-ft 9.9-in)
Launch Weight: 120-kg (264-lb)
Speed: Mach 3.5
Range: around 15-km
Guidance: IR homing head

Rafael Python 5

In June 2003, Rafael unveiled the Python 5 short-range IIR guided air-to-air missile. This missile, described as a fifth generation weapon, is based on the proven Python 4 aerodynamic configuration. The new missile provides the pilot of the launch aircraft with full-sphere launch capability when engaging a target. The missile has a claimed operating envelope described as from very short-range to BVR, featuring increased kill probability and resistance to countermeasures over its predecessor.

The Missile, which weighs in at 103.6 kg, has a length of 3096 mm, span of 64 cm and a diameter of 16 cm, has been integrated onto a number of aircraft types including IDF Boeing F-15 Eagle variants, IDF F-16 Fighting Falcon variants, IAI (Israeli Aircraft Industries) Kfir fighters and the Northrop Grumman F-5.

A Python 4 IR guided missile is carried on the starboard side wingtip station on a Saab JAS 39 Gripen fourth generation fighter. The aircraft is also carrying a mock-up of the European consortium Meteor. Saab

Above: The Taiwan developed Tien Chien 1 (TC-1) short-range infrared guided air to air missile looks more or less like a clone of the Raytheon AIM-9L Sidewinder. The Taiwanese missile arms Taiwan's Northrop F-5E and indigenous AIDC FCK-1 Ching Kou fighters.

The Python 5 introduces a number of new advanced technologies including a new duel-waveband imaging seeker, INS (Inertial Navigation System), advanced computer architecture, sophisticated IRCCM and advanced flight control algorithms. The rocket motor, warhead, proximity fuse and INS of the Python 4 are retained.

Python 5's full-sphere competence is achieved by a combination of LOAL (Lock-On After Launch) and acquisition and tracking capabilities. The duel-waveband FPA (Focal Plane Array) seeker and sophisticated algorithms allow acquisition of small targets emitting very low IR signatures in look-down, adverse and environments obscured by cloud.

A Tien Chien 1 IR guided air-to-air missile is launched from the starboard wingtip missile launch rail of a Taiwanese Northrop F-5E during development testing.

By mid-2003, development and captive carry tests of the missile had been conducted and the missile was expected to enter service with the Israeli Air Force, complementing the active radar guided Derby air-to-air missile.

Tien Chien I

The Tien Chien I (Sky Sword I) has been developed by Taiwan's CSIST as an infrared guided short-range air to air missile for service on the Taiwanese Air Force fighter fleets. Development of the missile commenced in the mid-1980's and the first air launch was conducted in April 1986 against a Beech 1089 target drone. Initial service entry was on Taiwanese Air Force Northrop F-5E and indigenous AIDC Ching-Kuo IDF (Indigenous Defence Fighter). The weapon is also to be cleared for use by Taiwan's Lockheed Martin F-16A/B Block 20 and Dassault Mirage 2000-5 multi-role fighters. Tien Chien I entered full-scale production in 1991 with the missile entering service with the Taiwanese - RCAF (Republic of China Air Force) in 1993.

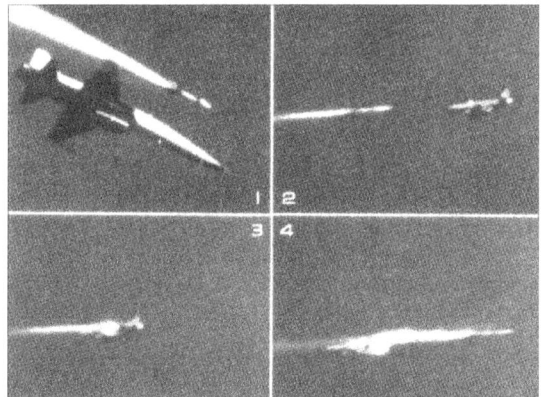

Top right: A sequence of stills from a video of a TC-1 missile launch from an F-5E to impact with the target drone. Above: A Tien Chien 1 missile is launched from the port side wingtip station of an AIDC Ching Kou fighter.

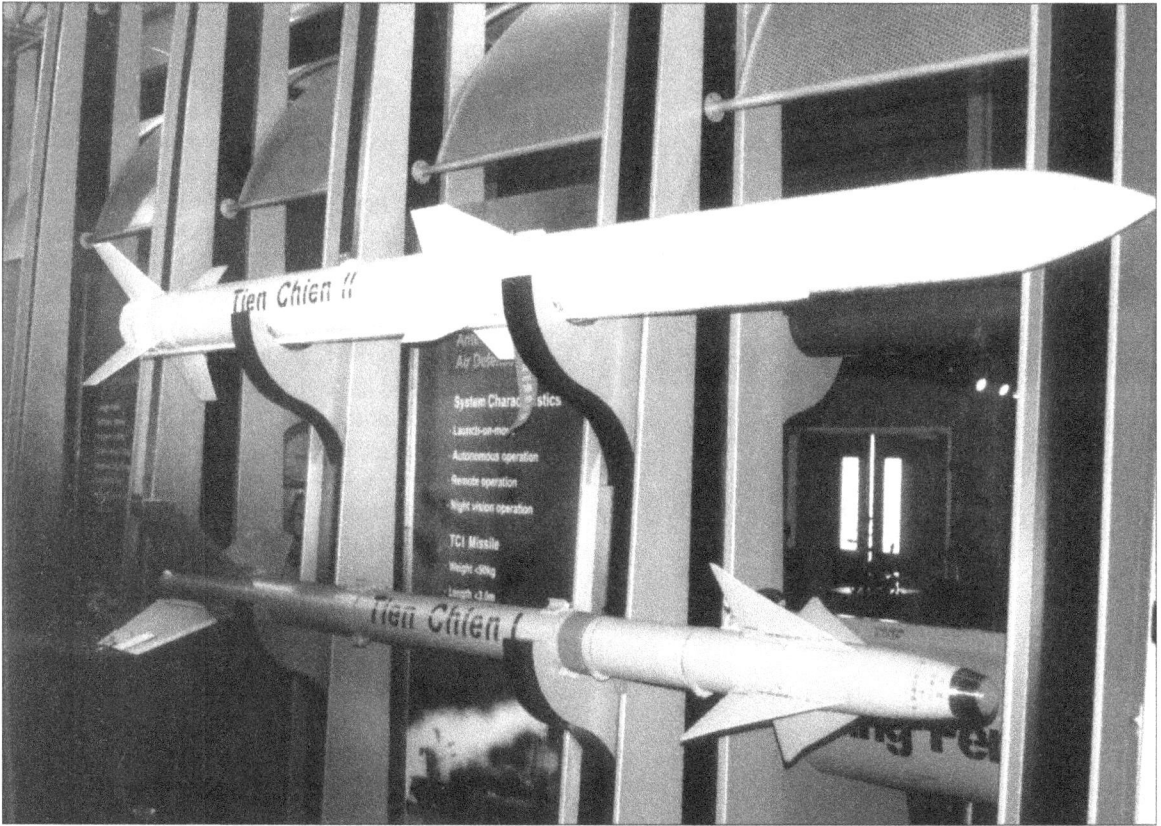

Previous page top and bottom: The TC-1 (bottom of both photographs) and TC-2 (top of both photographs) air to air missiles show obvious external similarities with the US AIM-120 and AIM-9L missiles respectively. However, it is assumed that both missiles are equipped with indigenous seeker-heads. Author

Top: This model of a Ching Kou fighter shows four TC-1 missiles being carried on the wingtip and outboard under wing stations and four TC-2 missiles being carried on the inboard under wing and fuselage centreline stations. Author Centre: A TC-2 missile is launched from a Ching Kou development aircraft. Above: A Ching Kou banks to starboard showing the centreline station carriage of TC-2 missiles.

The Tien Chien I employs a sensitive infrared seeker featuring all-aspect engagement capability. The missile can be used in fire and forget and slave-by-radar modes. Wind tunnel testing and CFD (Computational Fluid Dynamics) conducted in the 1980's showed that a canard controlled aerodynamic configuration with a roller on similar to the AIM-9L Sidewinder, which the Tien Chien I closely resembles was the most efficient design configuration. The aerodynamic design resulted in a missile with good manoeuvrability capable of engaging manoeuvring targets.

The missile is equipped with a modern GCU (Guidance Control Unit) and destroys its target with a high explosive warhead, which can be detonated by a contact or active laser proximity fuse. The missile is propelled by a low smoke solid-propellant rocket motor.

Tien Chien II

While work was progressing on the Tien Chien I short-range IR guided missile, CSIST began work on a medium-range active-radar guided air-to-air missile known as the Tien Chien II (Sky Sword II) in the 1980's. This missile was developed with all-weather, all-aspect; shoot up/shoot down capability and for operations in a dense ECCM environment. The RCAF Ching-Kuo IDF can carry two Tien Chien II missiles in tandem under the centre fuselage.

The Tien Chien II configuration features a tail-controlled aerodynamic design, bestowing good manoeuvrability. The missile is made up of four main sections consisting of a motor section, a warhead section, an actuator section and a high-reliability electronic section. This latter section consists of computer, inertial measurement unit, power supply and firing circuitry.

Integration with the Ching-Kuo fighter, including air launches, was completed in July 1994. That same year the missile entered production with service entry with the RCAF occurring in 1996.

Tien Chien II
Length: 3.60-m
Body diameter: 190-mm
Wingspan: 0.62-m
Launch weight: 183-kg
Warhead: 22-kg high explosive blast/fragmentation
Fuse: active-radar
Guidance: active-radar
Range: 60-km

The AAM-3 is the standard short-range IR guided air-to-air armament for the Mitsubishi F-2 support fighter. An AAM-3 is seen on the wingtip launch station of an F-2 model. Author

Mitsubishi Electric Corporation AAM-3

Japan began development of indigenous air-to-air missiles in the mid-1950's resulting in the AAM-1 and AAM-2 early generation IR guided air-to-air missiles. Development of the Type 90 (AAM-3) fourth generation IR guided air-to-air missile introduced much improved capability over the earlier weapons particularly in the field off manoeuvrability, turn-performance, IRCCM (Infrared Counter Counter Measures) and off-boresight engagement capability. The AAM-3 has been developed and fielded as a replacement for the AIM-9L Sidewinder in JASDF (Japan Air Self Defence Force) service. On-going development work may result in improvements to in-service and future variants.

Mitsubishi Electric Corporation AAM-4/4B

In the 1990's the JASDF was reliant on the SARH AIM-7M Sparrow and later purchased AIM-120 AMRAAM's as the primary BVR air to air armament for the services Boeing F-15J and Mitsubishi F-2 fighters. In the 1990's Mitsubishi pushed on with development of the XAAM-4 BVR air to air missile, which was based on the AIM-120. This weapon, which was ready for service by

the late 1990's and apparently in service from the early to mid-2000's, has a body made up of three main sections. The guidance section is at the front, the warhead located behind, followed by the propulsion system at the rear. Benefiting from Japanese expertise in advanced electronic systems, the AAM-4 has good ECCM resistance and is equipped with an extremely advanced active-radar seeker head. Several missiles can be fired at separate targets almost simultaneously.

Mitsubishi continued development of the weapon, which has now matured into the AAM4B, a much more capable weapon than the AIM-120 AMRAAM or the original AAM-4. The AAM-4 was designed as a replacement for the AIM-7 in JASDF service, while the AAM-4B is designed as a partial replacement for the services AIM-120 and AAM-4. The AAM-4B has been designed as a more capable weapon than its predecessors with on-board AESA (Active Electronically Scanned Array) radar housed in the missile nose allowing the launch aircraft to turn away much quicker after launch. This is due to the AESA being able to take-over the homing phase of the missiles flight at a far greater range than is possible with the AIM-120 or AAM-4. The less powerful radar incorporated in AMRAAM requires the launch aircraft, or a companion aircraft, to continue to illuminate the target for a longer period before the small active-homing radar can lock-on for the terminal phase, which inevitably exposes the illuminating aircraft to a higher threat environment than that encountered by a missile in the AAM-4B class.

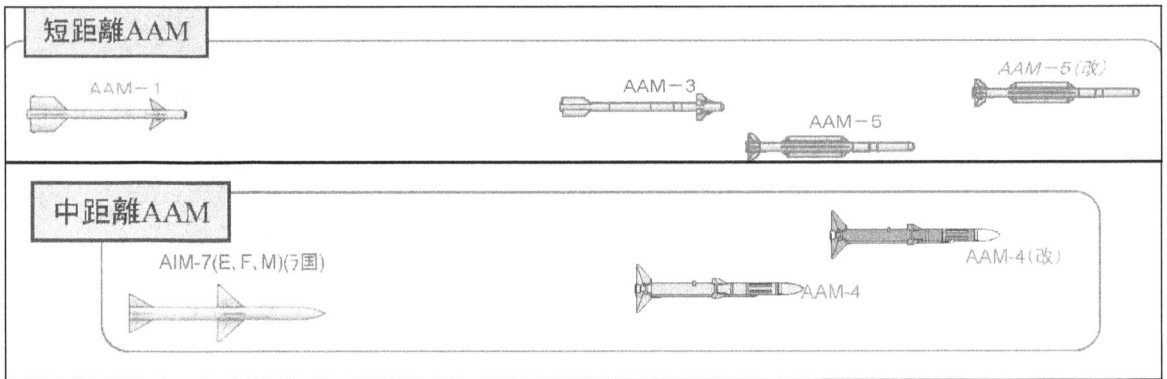

短距離AAM

AAM-1   AAM-3   AAM-5   AAM-5(改)

AAM-5

中距離AAM

AIM-7(E, F, M)(ア国)   AAM-4   AAM-4(改)

The AAM-4B has a weight of 230 kg (506 lb), with a diameter of 200 mm (8 in) and a length of 3.67 m (11.37 ft). Maximum effective range of the missile is stated as 120 km, well in excess of the range of the AIM-120 AMRAAM.

The AAM-4B is being procured to equip about 60 upgraded Mitsubishi F-2 fighters in service with the JASDF, although the weapon may well be introduced to Japans future fighter program. As well as being equipped with the AAM-4B, the radar of the F-2 will be upgraded from J/APG-1 to J/APG-2 standard allowing the fighter to take full advantage of the new capabilities of the AAM-4B, which with among other benefits has an increase in target detection range. The weapon was apparently the first air to air missile equipped with an AESA seeker head. As well as increasing the effective detection range of the missile, AESA bestows other improvements on the weapon such as a better performance against crossing targets according to the manufacturer.

In the case of the AAM-4 and the later AAM-4B, it has never been officially disclosed how many targets can be engaged simultaneously by the launch aircraft, with several missiles simultaneously being the mentioned.

A short-range surface to air missile variant is also being developed for Japanese naval surface combatants.

Mitsubishi Electric Corporation XAAM-5

The XAAM-5 short-range air to air missile is being developed by Japan as a successor to the AAM-3 short-range IR guided air-to-air missile currently in service with JASDF fighters. The XAM-5 will be a much more capable weapon, capable of meeting the threats and projected enemy countermeasures expected in the future.

The XAAM-5 will have extended lock-on range, enhanced IRCCM (Infrared Counter Countermeasures) and increased high-off boresight capability. The Mitsubishi F-2 support fighter and the upgraded Boeing F-15J Eagle and their successor will probably field the XAAM-5 missile should it enter service.

Glossary

AAAM	Advanced Air-to-Air Missile
AAM	Air-to-Air Missile
AEW	Airborne Early Warning
AI	Airborne Interception
AIM	Airborne Interception Missile
AMRAAM	Advanced Medium Range Air-to-Air Missile
ARC	Atlantic Research Corporation
ASRAAM	Advanced Short-Range Air-to-Air Missile
ATF	Advanced Tactical Fighter
AWACS	Airborne Warning and Control System
BITE	Built-In Test Equipment
BVR	Beyond Visual Range
BVRAAM	Beyond Visual Range Air-to-Air Missile
CFT	Conformal Fuel Tank
CSP	Capability Sustainment Program
CW	Continuous Wave
EADS	European Aeronautical Defence and Space
ECM	Electronic Counter Measures
ECCM	Electronic Counter Counter Measures
ERAAM	Extended Range Air-to-Air Missile
ESSM	Electronic Sensor Sensor Measures
F	Fighter
FA	Fighter Attack
FG	Fighter Ground Attack
FGR	Fighter Ground attack Reconnaissance
FMRAAM	Future Medium Range Air-to-Air Missile
FSO	Front Sector Optronics
GR	Ground attack Reconnaissance
HMS	Helmet Mounted Sight
HMSS	Helmet Mounted Sighting System
IDS	Interdictor Strike
IIR	Imaging Infrared
INS	Inertial Navigation System
IR	InfraRed
IRCM	InfraRed Counter Measures
IRCCM	Infrared Counter Countermeasures
IRIAF	Islamic Republic Iranian Air Force

IRIS-T	InfraRed Imaging Seeker-Tail control
IRST	InfraRed Search and Track
JHMCS	Joint Helmet Mounted Cueing System
LMTAS	Lockheed Martin Tactical Aircraft Systems
MBDA	Matra BAe Dynamics Alenia
MICA	*Missile d' Interception de Combat et d' Autodefence* (Dogfight Self-Defence/Interception Missile)
MLU	Mid-Life Update
MoD	Ministry of Defence
MRAAM	Medium Range Air-to-Air Missile
NATO	North Atlantic Treaty Organisation
NAWC-WD	Naval Air Warfare Centre-Weapons Division
RAAF	Royal Australian Air Force
RAF	Royal Air Force
RCAF	Republic of China Air Force
RFI	Request For Information
RSAF	Royal Saudi Air Force
SAM	Surface to Air Missile
SARH	Semi-Active Radar Homing
SRA	Staff Requirement Air
SRAAM	Short-Range Air-to-Air Missile
STOVL	Short-Take-Off and Vertical Landing
UAE	United Arab Emirates
UAV	Uninhabited Air Vehicle
UK	United Kingdom
US	United States
USAF	United States Air Force
USANG	United States Air National Guard
USMC	United States Marine Corp
USN	United States Navy
VFDR	Variable Flow Ducted Rocket

ABOUT THE AUTHOR

Hugh, a historian and Author, has published in excess of thirty books; non-fiction and fiction, writing under his own name as well as utilising two different Pseudonyms. He has also written for several international magazines, while his work has been used as reference for many other projects ranging from the Aviation industry, international news corporations, film media to encyclopedias and the computer gaming industry. He currently resides in his native Scotland

Other titles by the Author include

British Battlecruisers of World War 1 – Operational Log, July 1914-June 1915
Hurricane IIB Combat Log – 151 Wing RAF North Russia 1941
RAF Meteor Jet Fighters in World War II, an Operational Log
Typhoon IA/B Combat Log - Operation Jubilee August 1942
Eurofighter Typhoon – Storm over Europe
Tornado F.2/F.3 Air Defence Variant
Boeing X-36 – Tailless Agility Flight Research Aircraft
X-32 – The Boeing Joint Strike Fighter
X-35 – Progenitor to the F-35 Lightning II
X-45 Uninhabited Combat Air Vehicle
F-84 Thunderjet – Republic Thunder
USAF Jet Powered Fighters – XP-59-F-85
XF-92 – Convairs Arrow
The Battle Cruiser Fleet at Jutland
Saab Gripen, the Nordic Myth
American Teens
Dassault Rafale – The Gallic Squall
F/A-18E/F Super Hornet

CP

www.ingramcontent.com/pod-product-compliance
Lightning Source LLC
Chambersburg PA
CBHW051118200326

41518CB00016B/2546